RESILIENCE AND URBAN RISK MANAGEMENT

PROCEEDINGS OF THE CONFERENCE 'HOW THE CONCEPT OF RESILIENCE IS ABLE TO IMPROVE URBAN RISK MANAGEMENT? A TEMPORAL AND A SPATIAL ANALYSIS', PARIS, FRANCE, 3–4 NOVEMBER 2011

Resilience and Urban Risk Management

Editors

Damien Serre
Université Paris-Est, Ecole des Ingénieurs de la Ville de Paris, Paris, France

Bruno Barroca
*Université Paris Est, LEESU (Laboratoire Eau Environnement et Systèmes Urbains)—
Département Génie Urbain, Champs sur Marne, Marne la Vallée, France*

Richard Laganier
Université Paris Diderot, Sorbonne Paris Cite, Paris, France

CRC Press
Taylor & Francis Group
Boca Raton   London   New York   Leiden

CRC Press is an imprint of the
Taylor & Francis Group, an **informa** business

A BALKEMA BOOK

CRC Press/Balkema is an imprint of the Taylor & Francis Group, an informa business

© 2013 Taylor & Francis Group, London, UK

Typeset by V Publishing Solutions Pvt Ltd., Chennai, India

Published by: CRC Press/Balkema
P.O. Box 447, 2300 AK Leiden, The Netherlands
e-mail: Pub.NL@taylorandfrancis.com
www.crcpress.com – www.taylorandfrancis.com

ISBN: 978-0-415-62147-2 (Hbk)
ISBN: 978-0-203-07282-0 (eBook)

Table of contents

Resilience and Urban Risk Management – Serre, Barroca & Laganier (eds)
© 2013 Taylor & Francis Group, London, ISBN 978-0-415-62147-2

Preface

Towns concentrate activities, possessions and persons. When events such as an earthquake, flooding, an industrial accident or a combination of several hazards occur, the way towns operate is generally totally disrupted. Depending on the gravity of the event, the effects of the crisis may be felt on a much wide scale. Therefore, assessing a town's resilience in the face of risks is an essential step for being able to instil resilience policies afterwards.

Today, several reasons make urban risks complicated to manage. Since 2007, half the world's population live in urban environments. After reaching this threshold, it is now expected that the total population of people living in towns and cities will double in the next thirty years. This rate of increase is equal to building one new town with a million inhabitants every week (www.floodresiliencegroup.org, 2009). This rapid increase alone demonstrates the issues at stake in mastering urban risks.

Rapid urbanisation is generally accompanied by important urban sprawl. Urban sprawl generates risks by itself, as, on the one hand, it ends up by towns being built in zones where natural hazards are more likely to occur and, on the other, urban technical networks, too quickly designed and often under-dimensioned, exacerbate the effects of the hazard by their chain-reaction dysfunctions. Moreover, this rapid urbanisation is accompanied by climatic dynamics with extreme phenomena, which will also have an effect. IPCC experts point out that climate change will also generate a certain number of uncertainties depending on the temperature increase scenarios being considered.

However, irrespective of the scenario that is chosen, consequences are expected to occur both on the frequency and severity of rainfalls and this will also be the case for heat-waves and drought periods like the ones that took place in 2003. Predictive hazard models are produced on the basis of this uncertain, incomplete and inaccurate data.

In this context where the world is becoming increasingly uncertain, new strategies need to be developed for managing risks in order to anticipate scenarios that probabilistic models judge to be extreme or rare.

Moreover, independent of climate change, the growing concentration of possessions and persons in urban environments and, therefore, of potential vulnerability, makes devastating events very likely to occur in the coming years. For example, flood risks are liable to increase significantly as a result of the urbanization processes under way at present. The economic cost of flood risks throughout the world should reach a value of 100 billion euros per year by the end of the century. About 75% of all damage will be found in urban environments. For heat-wave and drought periods, cities, which often possess little greenery and therefore very little capacity to cool down during the night, will suffer directly from the negative effects of these heat-island phenomena.

Increased vulnerability results from these social phenomena (urban sprawl and urbanisation in dangerous zones) not only due to extreme events, but also to events considered to be current in days gone by.

Managing risks in cities has become complex: changing paradigms is becoming a necessity. New strategies must be envisaged for managing risks in urban environments. They must integrate a multitude of negative factors such as urban development and extreme climatic dynamics. This form of integrated management must act on several space and time scales.

The issue at stake most certainly concerns increasing a city's resilience in the face of risks, but also designing new urban neighbourhoods or new towns, where all these issues are taken into account: Certain town planners advocate transforming hazards, seen as negative events, into urban opportunities.

This means designing towns that are adapted to risks.

The purpose of this work, the fruit of an international seminar organized in Paris on November 3rd and 4th, 2011, is to put the resilience concept as applied to urban risk management into question. As a result, the major hypothesis emitted consists of considering the resilience concept as an opportunity for changing paradigms in order to increase the efficiency of urban risk management on the one hand, and to have urban designs better suited to the risks expected to occur over the coming years, and thereby drastically changing all the established standards and dogma.

It is organized in the form of two complementary sections:

- The temporal approach to resilience. In this first section, we are concerned about rethinking territories' capacity of adapting themselves today in the face of natural hazards, and designing networks and cities capable of resisting the menaces to which they will be exposed by relocating today's dynamics in a longer time perspective. Therefore, drawing lessons from the past through close examination of major catastrophes in history and the answers given to them by societies forms an initial key for assessing resilience factors in urban systems. Rethinking the weight of legacies from the past, more especially related to old methods of risk management linked with nature, offers a complementary perspective, which, together with the resilience concept, will lead to re-questioning sciences and technologies working on the problems of risk management and urban territories. In this way, we will try to see, in the light of the analysis of major events in history and evolutions in sciences and technologies, whether it is possible to integrate the exceptional, rare nature of major catastrophes into the way tomorrow's cities should be built.
- The second section is devoted to the "spatial" dimension. In what urban, technical, morphological, programmatic and social configurations could the resilience of an urban territory subjected to risks be conceived? This means "living life in the city with" the risk, rather than trying to protect ourselves by separating risk territories from urban territories. Researchers who have taken part in overseas projects are called upon and they propose a novel approach with a "neighbourhood" scale of study and with a "resilience" reference concept. Lastly, we are not concerned in this case by proposing an assessment of resilience, but by trying to define actions to be implemented for reaching a target resilience level. Then, the question of large territories comes to light through studies of organisational resilience and of urban networks together with trans-territorial forms of solidarity. Tools and methods recognized or tested in major international cities and their suburbs question city governance in a multi-scale and multi-player report and by action with local populations.

This multidisciplinary work is addressed to both researchers and backers, architects, town planners, and engineers who would like to see an urban project put into relation with the numerous facets of the urban resilience concept via concrete applications and methodological or historical reflections. In this way, the work recounts the latest progress made in terms of designing resilient towns, and it identifies leads to be explored for attaining the objective of systematically integrating risks into urban environments The work is a contribution to designing cities of tomorrow and to creating a new form of risk management from the resilience concept that does not ignore what already exists, allowing the integration of urban development and risk management.

Damien Serre, Bruno Barroca & Richard Laganier

About the editors

Damien SERRE, HDR, Professor Assistant at the Paris-Est University, EIVP, in charge of the "urban resilience" research section. The final objective of his research is to formalize knowledge useful for decision-making and helping in designing towns that are resilient in the face of risks. His research is trans-disciplinary and at the service of the city.

Bruno BARROCA, Architect and Professor Assistant in Urban Engineering at the Paris-Est University, a member of the urban engineering team of the LEESU laboratory (Water, Environment and Urban Systems Laboratory). His research establishes links between geography, town planning and regional development. Applications cover assessment of urban vulnerability and integration of resilience objectives in urban projects located on territories subject to natural and technological risks.

Richard LAGANIER, Professor in Geography at the Université Paris 7 Paris Diderot, Sorbonne Paris Cité, the PRODIG laboratory (Centre of Research for Organization and Distribution of Geographical Information). His research activities cover the study of relationships between risks linked with water and territories and analysis of the conditions needed for developing resilience. He is the author or co-author of a large number of works on hydrological extremes and their management.

Sponsors

Resilience and Urban Risk Management – Serre, Barroca & Laganier (eds)
© 2013 Taylor & Francis Group, London, ISBN 978-0-415-62147-2

Resiliencery Vulnerability notion—looking in another direction in order to study risks and disasters

D. Provitolo
UMR Géoazur, CNRS, Sophia Antipolis, Valbonne, France

ABSTRACT: Human societies and territories have always been confronted with hazards and risks, which are occasionally the source of disasters. Over the last 30 years, research on risks and disasters has greatly increased and focused on the vulnerability and resilience of societies. Resilience is often presented as an antonym of vulnerability, but in fact, the relationships are somewhat more complex. This paper adresses one of the challenges for research today—that of understanding the relationships established between the concepts of vulnerability and resilience. In order to understand these relationships, we propose a conceptual model of Resiliencery Vulnerability. This model is based on current knowledge of the risk and disaster domain. But although resilience is often presented as a sort of ultimate aim, we see things differently. The notion of Resiliencery Vulnerability expresses the idea that vulnerability is not necessarily a concept with a negative connotation and resilience a concept with a positive connotation.

The Resiliencery Vulnerability is a new notion (notion which may seem surprising because it is a neologism) introducing the idea that resilience may be contingent and not necessary, have a negative effect and that vulnerability can have a positive effect when change leads to a positive transformation. This notion applies to the study of risks and disasters. Risks and disasters have been evoked in a great deal of research into the three concepts hazard, vulnerability and resilience or the analysis of the vulnerability of populations, of geographical areas and of territories occupied by man (for example Adger, 2006, Becerra & Peltier (eds), 2009, Beck et al. 2012, Combe 2007, Cutter 1996, 2000, Daudé et al. 2009, Dauphiné 2003, D'Ercole et al. 1994, D'Ercole & Metzger 2009, Füssel 2007, Gilbert 2006, Gunderson & Holling 2002, Janssen et al. 2006, Kasperson et al. 2005, Kervern 1995, Leone 2008, Lhomme et al. 2010, Luers 2005, Magnan 2009, November 2006, Pigeon 2005, Provitolo 2007, Reghezza-Zitt 2009, Revet 2009, Rufat 2008, Ruin 2010, Uitto 1998, Veyret 2004, Weichselgartner 2001, Wisner 1999, 2004). For the last twenty years, conceptually speaking, the terms vulnerability and resilience have been scientific interest. The challenges of Vulnerability and Resilience Research are to understand the relationships woven between resilience and vulnerability and to question the idea that vulnerability is necessarily a concept with a negative connotation and resilience a concept with a positive connotation.

The aim of this article is to propose a systemic model in order to analyse the vulnerability and resilience of different systems, socio-economic systems, territorial systems in situations of risk or disaster. Two lines of research are developed. In the first part of the article, we are going to show the evolution and multiplicity of meanings of the concepts of vulnerability and resilience by analyzing disciplinary and cross-disciplinary research into risk, disaster and climate change. This analysis is used in the second part of the article to question the trio - Vulnerability-Event-Resilience and to understand the relationships that exist between resilience and vulnerability. In order to take account these relationships, we are going to present a systemic risk model based on the notion of Resiliencery Vulnerability. The idea is to help societies to anticipate and understand disasters that are becoming more and more complex and systemic, as we now see with the Japan disaster in 2011.

1 VULNERABILITY AND RESILIENCE, A RAINBOW OF MEANINGS

Multiple definitions of vulnerability and resilience exist within the literature. They have no broad accepted single definition (Klein et al. 2003, Manyena 2006). Although vulnerability and resilience can apparently only be understood in relation to each other (Van der Leeuw & Aschan Leygonie 2001, Gallopin 2003, 2006), these two concepts remain polysemic and their multiple definitions tend to obscure their meaning. That is why, we are going to present the different meanings of the concepts of vulnerability and resilience by looking at the scientific context in which they arose. We are also going to highlight the relationships between vulnerability and resilience and to show if there is some overlap in these concepts.

1.1 *Vulnerability in the spotlight*

If risk has long been restricted to the study of hazard, a change became apparent in the nineteen fifties with the appearance of the notion of vulnerability. Since then this notion of vulnerability has become ever more affirmed, enhanced and complex. Hazard alone can no longer explain the occurrence of a risk and the consequences of a disaster (Veyret 2004).

But the recent character of research in the domain of vulnerability and the interdisciplinary aspect of the subject matter make it a notion that has many meanings. The notion of vulnerability has not yet been fully explored (Gilbert 2009), in particular under the issues of climate change and gradual risks (Brooks 2003, Magnan 2009).

Behind the concept of vulnerability lie three approaches. The first examines vulnerability from the perspective of damage assessment (human, material, species, heritage), the second looks at vulnerability in terms of the response capacities of a system subjected to a hazard (e.g. human societies experiencing a disturbance), while the third identifies strategic areas and territorial stakes. In the first case, vulnerability is the outcome of the impact of the hazard upon the system. The scientific community talk about biophysical vulnerability. In the second and the third cases, social and territorial vulnerabilities are identified and evaluated independently of the hazard.

Based on the different approaches -biophysical, social (Adger 1999, Adger & Kelly 1999) and territorial vulnerabilities (D'Ercole & Metzger 2009), we would like to try and show that beyond the bounds of these concepts, vulnerability has opened the way to the concept of resilience.

The term biophysical vulnerability suggests a physical component, a biological component for a natural system or a social component for a human system (Veyret et al. 2004). Three factors influence the biophysical vulnerability of a system: the exposure of the elements to the event, their resistance and their sensitivity (Adger 2006). The level of exposure is still defined today by the nature, magnitude and frequency of the hazard and the proximity of societies and territories to the hazard area (Alexander 1993, Heyman et al. 1991). Resistance is the possibility for a system to absorb or counter the effects of a disturbance without sustaining damage. This may involve the physical resistance of infrastructures (floodwater hitting a dyke) or the physical or mental resistance of a person or group of individuals. Sensitivity is the degree to which people and places can be harmed (Adger 2006), more widely the sensitivity is understood as the damage that the system is liable to incur. Biophysical vulnerability is therefore based on estimation of damage in terms of losses, and more rarely in terms of gains. And this approach underestimates the social dimension of vulnerability.

Social vulnerability is considered as the capability of living systems to anticipate, cope with, manage the event and recover. It is a state that exists within a system before it is confronted with a hazard (Adger 1999, Adger & Kelly 1999). As such, social vulnerability is inherent to a system (Brooks 2003) and depends neither on exposure to the disturbance nor on its intensity. Research has also been done into identifying the factors determining social vulnerability (Cutter 1996, 2003, Pigeon 2005) According to the authors, social vulnerability depends on the level of resources—income, capital, social networks—and the level of accessibility to credits and information (Blaikie et al. 1994, Wisner 1998, 1999, Cross 2001), cultural and institutional factors (Bolay et al. (eds) 2012, Thouret & D'Ercole 1996, Ouallet 2009), technical and organizational factors. It thus involves a qualitative approach that focuses on the identification of vulnerability factors intrinsic to the social system studied.

But the social vulnerability approach does not include the functioning of territories, the strategic elements, such as networks of metropolitan areas, « lifeline ». To overcome this limit, other authors have proposed the concept of territorial vulnerability (Bonnet 2002, Cutter et al. 2003, D'Ercole & Metzger 2009).

The territorial vulnerability aims to identify, characterize and prioritize areas from which vulnerability is generated and disseminated in a territory. It is a question of identifing both fragile areas, likely to know of serious damage in case of disaster, but also the places likely to spread their vulnerability in a territory. The vulnerability is linked to territorial stakes and not to the exposure to the hazard.

The humanities and social sciences and interdisciplinary research groups develop the concept of vulnerability showing that the three approaches above are not conflicting. Vulnerability has two sides to it—it is both intrinsic to the system but it also varies with the system's capacity to incur a disturbance, to absorb it and to resume normal functioning. Human societies, for example, are capable of adapting to the hazard, learning, and thus modifying their exposure to the hazard. The 2001 and 2007 reports of the Intergovernmental Panel on Climate Change (IPCC) followed the same line of thought by defining vulnerability as "the degree to which a system is susceptible to, and unable to cope with, adverse effects of climate change, including climate variability extremes. Vulnerability is a function of the character, the magnitude, and rate of climate change and variation to which a system is exposed, its sensitivity, and its adaptive capacity". The notion of vulnerability thus paves the way for resilience.

1.2 *Resilience, a shifting and disparate concept*

Resilience is a concept of physics that has been transferred to the social sciences via ecology. In the course of these transfers among sciences, the concept has diversified and come to have more than one meaning.

1.2.1 *From resilience in its initial disciplines …*

Resilience, a key concept for analyzing ecosystems, was for a long time the domain of ecology sciences. Holling (1973) defined it as the capacity of an ecosystem to integrate a disturbance without modifying its qualitative structure. Resilience thus expresses both a system's capacity to resist during the disturbance and its capacity to confront it, to recover and to regenerate (for example, the growth of woodland after a fire or storm). This definition is similar to that of social vulnerability (previously presented). This dual capacity to resist and to recover, without any change in structure, may indeed be applied to many subjects of study, be they populations, societies, or cities for example.

Then since 1980', two approaches have come into conflict. One of them—a resilient system is a stable system that is close to a permanent state of equilibrium (Pimm 1984). Resilience is the capacity of a system that incurs a brutal shock or continual pressure to sustain itself unchanged. It is measured by its resistance and the rapidity of return to equilibrium. This is known as Engineering Resilience. The second approach—a resilient system is one that maintains its essential functions and structures by moving through different states of equilibrium (stable and unstable). This is Ecosystem Resilience or Ecological Resilience.

This concept of Ecosystem Resilience is more suitable for the study of complex adaptive systems as it can go beyond the single-equilibrium paradigm. Indeed, a complex system is a system open to its environment (in the systemic meaning of the term) that can theoretically move towards an equilibrium, but that can also shift towards stable or unstable stationary solutions that are a long way from equilibrium. Of course, all complex systems are not adaptive, but all those involving living societies are. This is why research in the humanities and social sciences (Cutter et al. 2008, Dauphiné & Provitolo 2007, Gallopin 2006, Maret & Kadoul 2008, Vis et al. 2001) or in cross-disciplinary fields has also enhanced the concept of resilience.

3

In the humanities and social science, the concept of resilience has spread to a variety of different contexts. Without striving to be exhaustive, research has been done into the resilience of people (Cyrulnik 2002), of spatial and territorial systems (Aschan Leygonie 2000, International Strategy for Disaster Reduction [ISDR] 2005, 2009, International Human Dimensions Programme on Global Environmental Change [IHDP] 2003), and other research has been conducted into the resilience of socio-economic systems (Décamps 2007) and urban productive networks, while other research has investigated the links between ecological and social resilience (Adger 2000).

In the face of these disciplinary approaches to resilience, a new concept, that of systemic resilience, is making an informal appearance.

1.2.2 *... to systemic resilience*

Transdisciplinary research concerned with nature/society interactions, environmental and social pressures (Berkes & Folke 1998, Berkes et al. 2003, Gunderson & Holling 2002, Holling 2001, Kasperson et al. 1995, Klein et al. 1998, Klein et al. 2003, Resilience Alliance, Smit & Wandel 2006, Walker et al. 2004) are developed in Resilience Alliance.

The Resilience Alliance is a multi-disciplinary research group that studies the dynamics of complex adaptive systems and has particularly advanced the concept of resilience through the development of the idea of systemic resilience. This applies to all physical and social systems. Systemic resilience goes beyond the mere idea of resistance to change and preservation of existing structures preferring the ideas of system renewal, re-organisation and emergence of new trajectories.

Systemic resilience is interpreted by stylized models of the adaptive cycle (Walker et al. 2004) and of panarchy (Gunderson & Holling 2002). These models focus on the different trajectories following by a system. The panarchy model is a stylized representation for understanding multilevel transformations in natural and human systems. It is a model of transition between different steady states: connections between levels are made via two trajectories: one is "revolt" representing the transition from the collapse phase to the conservation phase; the second, the "remember" trajectory is the direct trajectory from the conservation phase to the re-organisation phase. Within the model of panarchy, "structures and processes are linked across scales" (Cutter et al. 2008). The multiscale approach of this heuristic model implies that the disappearance of a subsystem can reveal the resilience capacity of a metasystem.

Systemic resilience is therefore a concept well adapted to the management of risks with a physical and social dimension (Van der Leeuw & Aschan Leygonie 2001). But here again the definition is not clearly distinguished from that of vulnerability. To study the vulnerability of a population or system is also to analyze its capacity to recover from a situation, to recuperate. And yet the renewal of a system, its re-organisation, or even the emergence of new trajectories (all notions related to the concept of resilience) are properties of recuperation (a property related to the idea of vulnerability). An area of overlap arises between the concepts of vulnerability, resilience and their determinants (We mean exposure, resistance, sensitivity, adaptive capacity, adaptation, etc).

That is why, in the second part of this article, we are going to question the trio of Vulnerability-Event-Resilience and to present a systemic risk model based on the notion of Resiliencery Vulnerability.

2 A SYSTEMIC RISK MODEL BASED ON THE NOTION OF RESILIENCERY VULNERABILITY

From a fragmented body of literature presented above, from existing vulnerability and resilience models (Ashley & Carney 1999, Carter et al. 1994, Cutter 2008, Décamps 2007, Gunderson & Holling 2002, Kasperson & Kasperson 2001, Lim & Spanger-Siegfried 2005, Winograd 2004), we are going to present a model for analysing the Resiliencery Vulnerability of systems in the face of risk and disaster.

2.1 Resiliencery Vulnerability notion: Looking in another direction

Resilience is often presented as an antonym of vulnerability (Folke et al. 2002). Vulnerability is usually portrayed in negative terms as the susceptibility to be harmed (Adger 2006). We suggest a new conceptual framework in which the vulnerability and resilience of systems (e.g. social systems, economic systems, territorial systems) are understood as a continuum, as linked concepts. The Resiliencery Vulnerability notion that we propose provides an escape from the conception whereby vulnerability has a negative connotation concept and the concept of resilience is a positive response to a disturbance. This new notion challenges the idea that there are on one side vulnerable (and therefore weak) systems and on the other resilient systems (and therefore strong ones, capable of fighting back and recovering).

The Resiliencery Vulnerability notion opens up the field of the research, by introducing the idea that resilience may be contingent and not necessary, have a negative effect and that vulnerability can have a positive effect when change leads to a positive transformation. This is the power of vulnerability. For example, collective protection structures (e.g. dikes or flood barriers) are a form of societal resilience for a type of risk and an intensity. Nevertheless, the storm Xynthia which crossed Western Europe on 26–28 February 2010 showed that these defence works did not provide absolute protection, gave a false sense of security, and paradoxically increased the vulnerability of the population.

2.2 A systemic model for analyzing the Resiliencery Vulnerability of systems in the face of risk and disaster

We propose the Resiliencery Vulnerability model (Fig. 1) as a new conceptualization to study systems in the face of risk and disaster. The task is to provide a conceptual framework that can be used for a global analysis of the Resiliencery Vulnerability of elements brought into play by events. This conceptual framework's ambition is to be applicable to analyzing the

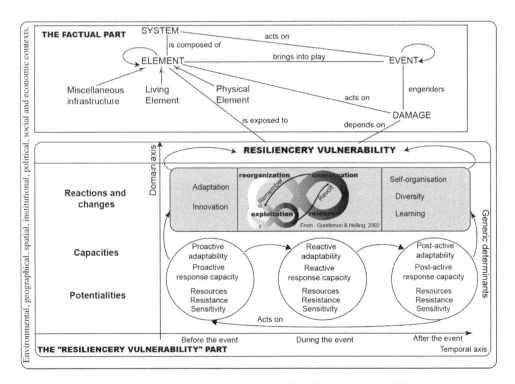

Figure 1.   A systemic Risk Model based on the notion of Resiliencery Vulnerability.

Resiliencery Vulnerability of territories exposed to various types of events, localized or diffuse, whether natural, industrial/technological or social.

The conceptual model distinguishes the factual part that describes the context (the elements and events) and depicting what happened (e.g. 30 people died on a bridge), from the Resiliencery Vulnerability part that describes the potentialities, capacities and reactions of a system to protect itself from risk or disaster for a complete temporal environment (before and during the event and after the disaster impact). This Resiliencery Vulnerability varies with time, from one country to another, depending on the environmental, geographical, spatial, institutional, political, social and economic contexts.

2.2.1 *Explanation of the factual part of the model*

To reduce disaster risks, we have to know which events societies must cope with and which elements are exposed to these events. This is why the factual part identifies the elements pertinent to the analysis of a system subjected to potentially disastrous events (Provitolo et al. 2009a) and the exposure of these elements to Resiliencery Vulnerability. The Element term generalizes the Living Element, Physical Element, and Miscellaneous Infrastructure classes that appeared to us to be the relevant categories to be distinguished in the case of risk and disaster. To define the terms previously mentioned:

- the Living Element includes all human beings and natural populations such as fauna and flora;
- the Physical Element corresponds to the description of the Earth's surface (oceanography, hydrography, pedology, relief, etc.) and is not directly concerned with human activities;
- the Sundry Infrastructures encompass built areas, facilities, networks, etc.

An event brings into play elements. A single event may bring several elements into play and any one element may be subject to several separate events. The relationship "brings into play" allows us to link Event and Element.

This relationship indicates that the element is exposed to the event, but that exposure does not presuppose its vulnerability. In this way elements may intervene as catalysts of the event; elements are not necessarily impacted. Element is in fact nothing other than the "matter" of which event is part.

An event can also cause other events. Events are thus linked together by causal relationships that are expressed *via* elements. This causal chain corresponds to domino effects. Lastly, the event can engender damage of different kinds and in variable amounts.

An event may be an actual (disaster), probable or potential phenomenon (risk), i.e. a threat of destruction or disruption (Perry 2006). According to the scale or magnitude of the event, according to the actions or inactions of an organization, according to the damages, an event can be classified in terms of risks, crises, disasters or catastrophes. For a literature review on the definitions of these terms see: Rodriguez et al. 2006, Faulkner 2001.

In a vulnerability context, Cutter (2005) argued that the issue is not disasters as events but instead human "vulnerability (and resilience) to environmental threats and extreme events". That is why this model dissociates the elements from their Resiliencery Vulnerability.

2.2.2 *Explanation of the Resiliencery Vulnerability part of the model*

The two axes that compose the Resiliencery Vulnerability part of the model—the temporal axis and the domain axis. The temporal axis distinguishes three temporal environments: before, during and after the event. The domain axis describes three phases: the potentialities, capacities and reactions of a system to protect itself from an event in a complete temporal environment. This model makes a distinction between what is available, what can be done and what is actually done. The three phases (potentialities, capacities and reaction) are closely linked in the sense that reactions and changes are constrained by the capacities (i.e. what can be done) and potentialities (i.e. available resources). Capacities are actualized when they correspond to reactions of a system. For each temporality, we identifie generic determinants of Resiliencery Vulnerability, as opposed to specific determinants relevant to a particular context and type of hazard. We take account the interactions between these determinants.

Resiliencery Vulnerability is a dynamic process; it is thus part of what happens in the pre-event time (the disaster has not yet happened, it might occur and so preventive strategies should be put in place ultimately to avert it or at least to minimize it), in the actual course of the event and in the future (after the event). The behavior of these three temporal environments, that are mutually but not simultaneously influential, is made more complex by the emergence of feedback loops. In the terms used by Thiétard (2000), "yesterday's actions are at the source of today's reactions which lead in their turn to new actions tomorrow". In relation to risk and disaster, yesterday's actions are those put in place within the risk system, especially through preventive measures and the regional development policy. Those actions of yesterday will act upon the course taken by the disaster and thus entail reactions during the event. Then, when the event is over, a post-disaster analysis, better known as an experience feedback, allows us to identify and analyze the failings and errors made in risk prevention and disaster management. De-briefings were conducted, for example further to the black ice that brought Montreal to a halt in January 1998 or to cyclone Katrina that hit New Orleans in 2005. New strategies to combat the recurrence of a possible disaster are then put in place.

Potentiality of a system correspond to the available resources, to the capital (economic, technological, social, human, knowledge …). These resources allow us to assess the resistance and sensitivity of a system (see definitions of resistance and sensitivity § 1.1). Space technologies (satellite remote sensing, GIS etc.) are for example technological resources used in disaster management. But, there is sometimes a hiatus between the resources available before and during the event (for example absence of satellite data for developing countries with no satellite of their own), and those actually available after the event.

The capacity phase allows the system to benefit from these potentialities and to implement them. The capacity to implement a system's potentialities is effected through the system's capacity for response and its adaptability (or adaptive capacity). These properties are described as being proactive, reactive, or post-active depending on their timing relative to the unfolding of an event (before, during, or after). According to some researchers adaptive capacity is defined as the ability of a system to adjust to change, moderate the effects and cope with a disturbance (Burton et al. 2002, Brooks et al. 2005). Adaptability (or adaptive capacity) represents a potential rather than actual adaptation (Brooks 2003). Adaptability does not predict what types of adjustments will occur, but gives an indication of the ability to adapt to the disturbance.

The adaptive capacity will act on the adaptation process and on the result, i.e. the system adapts itself. Generally speaking, adaptability is presented as a factor that increases the resilience of a system. But high adaptability can unintentionally lead to a loss of resilience in three cases (Walker et al. 2006):

− multiscale, multilevel: adaptability of certain groups of living elements to the detriment of others, adaptability at the microscale to the detriment of the macroscale,
− adaptability of the system for a specific shock may entail lower general resilience in the event of unplanned-for shocks,
− loss of diversity of responses.

The reactions and forms of change reflect both the properties actually implemented with a system exposed to risk or disaster and the different trajectories that the system follows. Regarding the different trajectories of the system, we have in fact incorporated in our conceptual model the panarchy model of Gunderson and Holling (2002). This model is coupled with the properties that will determine the dynamics of a system and thus its trajectories.

Among the properties that determine the dynamics of a system affected by a disturbance, five are often cited in the literature: diversity, self-organization, learning and adaptation (Adger et al. 2005, Klein et al. 2003, Folke 2006), innovation. In ecology, loss of biodiversity is considered to be a factor that reduces the ecosystem's resilience; the same is true of economies based on a single activity. Moreover, self-organizing systems have a greater capacity for restoration because the "functions" of the damaged parts are taken in hand by the other elements. However, Walker & Salt (2006) distinguish between "functional" and "response" diversity. Face a disturbance, what is important is that the different element forming part of the same functional group each have different responses (Elmqvist et al. 2003). Self-organising systems are therefore generally resilient. Likewise, resilience depends on the capacity of the living system to adapt, which is the case of living societies thanks to learning.

Adaptation refers both to the process of adapting and to the outcome. Adaptation can also be considered as an aim, an aim that is not fixed but instead changes over time. The adaptation is logically a proactive process for example by setting up risk prevention plan and/or crisis management. The adaptation of a system also reflects its capacity to avert an event (referred to as adaptation by learning) or to adjust to it. For example, in a society where the population is well prepared to react to a given type of event, panic behavior is less to be feared than in an inadequately or badly informed population (Provitolo et al. 2011). Lastly, innovation will promote the implementation of new practices, new actions to combat risk and modify Resiliencery Vulnerability.

2.3 *Validation of the Resiliencery Vulnerability systemic model*

The factual part of the model was validated from the analysis of two disasters: the Great Kanto earthquake (1923), (Provitolo et al., 2009b) and the 2011 Japan's triple disaster (Dubos-Paillard & Provitolo, in prep.). The validation of the factual part was performed from P. Hadfield's account "Sixty Seconds That Will Change the World: The Coming Tokyo Earthquake" for the analysis of the disaster of the Great Kanto Earthquake and from the gathering of information (textual and multimedia information) for the 2011 Japan's disaster.

In the first case, the use of the factual part "filter" for analyzing the Great Kanto Earthquake shows the need for a multi-scale approach to reconstruct the account as well as possible because each scale provides specific information about the event. This vision of an event at different scales has many advantages in terms of comparison: comparison of accounts pertaining to different catastrophes, but also comparison of events of the same scale localized in different places and part of the same catastrophe (Figs. 2–3). In the latter case, the use of the factual part "filter" for analyzing the 2011 Japan's disaster shows that many relationships exist between elements and events, between various events, and between events and damages, but also between damage and the elements they bear on. The reader interested in the validation may refer to the two above quoted references.

The "resiliencery vulnerability" part of the model is validated from the analysis of different catastrophic events. In this article, we are going to present different types of hiatus: those inside a domain, and those between domains.

In the domain of space technologies used in disaster management (satellite remote sensing, GIS etc.) there is sometimes a hiatus between the resources available before and during the

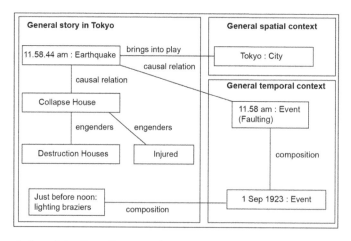

Figure 2.   Description of the Great Kanto earthquake at the scale of the city of Tokyo.

8

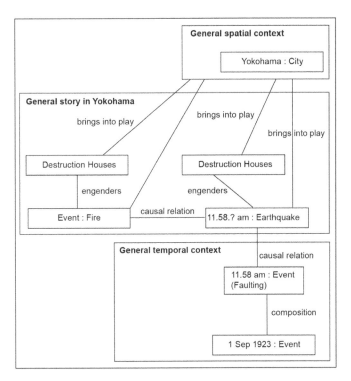

Figure 3.   Description of the Great Kanto earthquake at the scale of the city of Yokohama.

event (absence of satellite data for developing countries with no satellite of their own), and those actually available after the event. For example, all African countries do not have satellite data to monitor and manage disaster risks, particularly in the field of the prediction of and preparation against drought (*domain: potentialities before the event*). In order to provide equal data access, the International Charter on Space and Major Disasters supplies data free of charge to any country struck by disaster. The "satellite data" resource is then available following the event (*domain: potentialities after the event*). This charter does not help developing countries acquire data for mitigation, planning and preparation for disaster risks (*domain: potentialities before the event*). The Resiliencery Vulnerability model thus reveals a lack of coherence, of continuity in disaster event management due to the unavailability of satellite resources before, during and after the disaster.

Sometimes, there are gaps between two domains, those of potentialities and capacities. In developing countries or in countries subject to a high degree of corruption, numerous safety equipments are either obsolete or too small to manage the impacts of a disaster, to ensure water safety and quality. For example it is unusual for cities in developing countries to have abundant drinking water, even if it is a resource available before a disaster (*potentialities before the event*). This water resource most often disappears during a disaster (*lack of potentialities during and after the event*), thus altering the capacity for response. The resource is available before the event, but it cannot be used during and after the event (*domain: capacities during and after the event*) because it is no longer available (*domain: potentialities during and after the event*). Resiliencery Vulnerability of populations is then changed. For example, the cholera epidemic occurred just after the earthquake in Haiti (2010). And the Xynthia storm in France (2010) showed that, despite the existence of disaster prevention system (*potentialities domain*), the early warning system did not work (*potentialities during the event*). The resource composed of prevention system was not used efficiently (*capacities during the event*). Finally, during the reconstruction phase, a significant part of financial aids provided by the international community (*potentialities after the event*) is sometimes diverted. Sometimes, these financial supports, while necessary, compete with the local economy (*capacities after the event*), and thus have a counterproductive effect. A massive injection of food during famine, can reduce the income of farmers who can no longer sell their products (*reactions and changes*), which greatly increases their vulnerability.

3 CONCLUSIONS AND NEXT STEPS

The review of the literature revealed that there exist many definitions and types of vulnerability and resilience. The definitions of concepts have important implications in the choice of frameworks, theories, methodologies and tools as well as in the results of vulnerability assessments and resilience of territories. This paper has tried to address one of the challenges for research today—that of understanding the relationships woven between the concepts of vulnerability and resilience. We have reconsidered the idea that vulnerability is necessarily a concept with a negative connotation and resilience a concept with a positive connotation. A systemic risk model based on the notion of Resiliencery Vulnerability has been presented and partially validated. This does not deny that systems can be fragile. On the contrary, it allows us to shed light on the potentialities, the capacities and reactions that enable systems to protect themselves against risks or disasters and to confront them. The idea is to help societies to anticipate and understand disasters that are becoming more and more complex and systemic, as we now see with the Japan disaster in 2011. This model is certainly imperfect. The next step will be to validate it totally on different systems (geographical areas, cities, ecosystems) confronted with actual or probable events.

ACKNOWLEDGMENTS

This work was in part sponsored by the Conseil Général des Alpes-Maritimes (CG06), (France) in the framework of the project "Vulnerability and Resilience of Mediterranean cities". We wish to thank the CG 06 for research subsidies.

REFERENCES

Alexander, D. 1993. *Natural Disasters.* New York: Chapman & Hall.
Adger, W.N. 1999. Social vulnerability to climate change and extremes in coastal Vietnam. *World Development* 27 (2): 249–269.
Adger, W.N. 2000. Social and ecological resilience: are they related? *Progress in Human Geography* 24 (3): *347–364.*
Adger, W.N. 2006. Vulnerability. *Global Environmental Change* 16: 268–281.
Adger, W.N. & Kelly, P.M. 1999. Social vulnerability to climate change and the architecture of entitlements. *Mitigation and Adaptation Strategies for Global Change* 4: 253–266.
Adger, W.N. et al. 2005. Social-ecological resilience to coastal disasters. *Science* 309 (5737): 1036–1039.
Ashan Leygonie, Ch. 2000. Vers une analyse de la résilience des systèmes spatiaux. *Espace Géographique* 1: 64–77.
Ashley, C. & Carney, D. 1999. Sustainable livelihoods: Lessons from early experience. London: Department for International Development DFID.
Becerra, S. & Peltier, A. (eds), 2009. *Risques et environnement: recherches interdisciplinaires sur la vulnérabilité des sociétés.* Paris: L'Harmattan.
Beck E. et al. 2012. Risk perception and social vulnerability to earthquakes in Grenoble (French Alps). *Journal of risk research* 16 (7).
Berdica, K. 2002. An introduction to road vulnerability: what has been done, is done and should be done? *Transport policy* 9 (2): 117–127.
Berkes, F. & Folke C. (eds), 1998. *Linking Social and Ecological Systems. Management Practices and Social Mechanisms for Building Resilience.* Cambridge: Cambridge University Press.
Berkes, F. et al. (eds), 2003. *Navigating Social-Ecological Systems: building resilience for Complexity and Change.* Cambridge: Cambridge University Press.

Bolay, J.C. et al. (eds). 2012. Technologies and Innovations for Development: Scientific Cooperation for a Sustainable Future. France: Springer-Verlag.

Bonnet, E. 2002. *Risques industriels: évaluation des vulnérabilités territoriales.* Thèse de doctorat Université du Havre, 341 p.

Brooks, N. 2003. *Vulnerability, Risk and Adaptation: A Conceptual Framework.* Tyndall Center for Climate Change Research, working paper 38.

Brooks, N. et al. 2005. *Global Environmental Change* 15: 151–163.

Burton, I. et al. 2002. From impacts assessment to adaptation priorities: the shaping of adaptation policy. *Climate Policy* 2 (2–3): 145–159.

Carter, T.L. et al. (eds.) 1994. Technical Guidelines for Assessing Climate Change Impacts and Adaptations. *Report of Working Group II of the Intergovernmental Panel on Climate Change.* London, UK and Tsukuba, Japan: University College London and Centre for Global Environmental Research.

Combe, C. 2007. *La ville endormie ? Le risque d'inondation à Lyon. Approche géohistorique et systémique du risque de crue en milieu urbain et périurbain.* Thèse de doctorat de géographie, Université Lumière Lyon 2.

Cross, J.A. 2001. Megacities and small towns: different perspectives on hazard vulnerability. *Global Environmental Change Part B: Environmental Hazards* 3 (2): 63–80.

Cutter, S.L. 1996. Vulnerability to environmental hazards. *Progress in Human Geography* 20 (4): 529–539.

Cutter, S.L. et al. 2000. Revealing the vulnerability of people and places: a case study of Georgetown country, South California. *Annals of the Association of American Geographers* 90 (4): 713–737.

Cutter, S.L. et al. 2003. Social vulnerability to environmental hazards. *Social Science* 84 (1): 242–261.

Cutter, S.L. 2005. Are we asking the right question? In R.W Perry & E.L. Quarantelli (eds), *What is a disaster: new answers to old question.* Philadelphia: Xlibris.

Cutter, S.L. et al. 2008. A place-based model for understanding community resilience to natural disasters. *Global Environmental Change* 18: 598–606.

Cyrulnik, B. 2002. *Un merveilleux malheur.* Paris: Odile Jacob.

Daudé, E. et al. 2009. Spatial risks and complex systems: methodological perspectives. In M.A. Aziz-Alaoui & C. Bertelle (eds), *From System Complexity to Emergent Properties.* Berlin, Heidelberg, New York: Springer, Series Complex System.

Dauphiné, A. 2003. *Risques et catastrophes: observer, spatialiser, comprendre, gérer.* Paris: A. Colin.

Dauphiné, A. & Provitolo, D. 2007. La résilience: un concept pour la gestion des risques. *Annales de Géographie* 654 (2): 115–125.

D'Ercole, R. et al. 1994. Les vulnérabilités des sociétés et des espaces urbanisés: concepts, typologie, modes d'analyse. *Revue de géographie alpine* LXXXII (4): 87–96.

D'Ercole, R. & Metzger, P. 2009. La vulnérabilité territoriale: une nouvelle approche des risques en milieu urbain. *Cybergeo,* URL: http://www.cybergeo.eu/index22022.html

Décamps, H. 2007. La vulnérabilité des systèmes socioécologiques aux événements extrêmes: exposition, sensibilité, résilience. *Natures Sciences Sociétés* 15: 48–52.

Dubos-Paillard E. & Provitolo D., in prep. Analysis of the 2011 Japan's triple disaster through an ontology of the domain of risk and disaster. *In Complex System Design and Management 12–14 December, Paris.*

Elmqvist, Th. et al. 2003. Response diversity, ecosystem change, and resilience. *Frontiers in Ecology and the Environment* 1 (9): 488–494.

Faulkner, B. 2001. Towards a framework for tourism disaster management. *Tourism Management* 22 (2): 135–147.

Folke, C. et al. 2002. Resilience and sustainable development: building adaptive capacity in a world of transformations. *Report for the Swedish Environmental Advisory Council.* Stockholm, Sweden: Ministry of the Environment.

Folke, C. 2006. Resilience: the emergence of a perspective for social-ecological systems analyses. *Global Environmental Change* 16 (3): 253–267.

Füssel, H.-M. 2007. Vulnerability: A Generally Applicable Conceptual Framework for Climate Change Research. *Global Environmental Change,* 17 (2): 155–167.

Gallopin, G.C. 2003. A systemic synthesis of the relations between vulnerability, hazard, exposure and impact, aimed at policy identification. In *Handbook for Estimating the Socio-Economic and Environmental Effects of Disasters.* Mexico: ECLAC, LC/MEX/G.S.

Gallopin, G.C. 2006. Linkages between vulnerability, resilience, and adaptative capacity. *Global Environmental Change* 16 (3): 293–303.

Gilbert, C. 2006. La vulnérabilité, une notion à explorer, *Dossier Pour la Science* 51: 116–120.

Gilbert, C. 2009. La vulnérabilité: une notion vulnérable ? In S. Becerra & A. Peltier (eds.), *Risques et environnement: recherches interdisciplinaires sur la vulnérabilité des sociétés.* Paris: L'Harmattan.

11

Gunderson, L. & Holling, C.S. (eds). 2002. *Panarchy: Understanding Transformations in Human and Natural Systems.* Washington, DC.: Island Press.

Heyman, B.N. et al. 1991. An assessment of worldwide disaster vulnerability. *Disaster Management* 4: 3–14.

Holling, C.S. 1973. Resilience and stability of ecological systems. *Ann. Rev. of Ecol. and Syst.* 4: 1–23.

Holling, C.S. 2001. Understanding the complexity of economic, ecological and social systems. *Ecosystems* 4: 390–405.

ISDR, 2005. *Hyogo Framework for Action 2005–2015: Building the Resilience of Nations and Communities to Disasters.* Extract from the final report of the World Conference on Disaster Reduction, http:www.unisdr.org/eng/hfa/hfa.htm

ISDR, 2009. *Global Assessment Report on Disaster Risk Reduction: Risk and Poverty in a Changing Climate.* http://www.preventionweb.net/english/hyogo/gar/report/

IHDP, 2003. *Resilience,* Newsletter 02. http://www.ihdp.unu.edu/file/get/7189

IPCC, 2001. *Climate change 2001: Impacts, Adaptation and Vulnerability.* Contribution of Working Group II to the IPCC Third Assessment Report. New York: Cambridge University Press.

IPCC, 2007. *Climate Change 2007: Climate Change Impacts, Adaptation and Vulnerability.* Cambridge: Cambridge University Press.

Janssen, M.A. & Ostrom, E. 2006. Resilience, vulnerability, and adaptation: a cross-cutting theme of the International Human Dimensions Programme on Global Environmental Change. *Global Environmental Change* 16: 237–239.

Kasperson, J.X. & Kasperson, R.E. 2001. International Workshop on Vulnerability and Global Environmental Change. SEI Risk and Vulnerability Program Report.

Kasperson, J.X. et al. (eds.). 1995. *Regions at Risk: Comparisons of Threatened Environments.* Tokyo: United Nations University Press.

Kasperson, R.E. et al. 2005. Vulnerable people and places. In R. Hassan et al. (eds.), *Ecosystems and Human Well-Being: Current State and Trends* Washington (D.C.): Island Press.

Kervern G.-Y. 1995. Éléments fondamentaux des Cindyniques. Paris: Economica.

Klein, R.J. et al. 1998. Resilience and Vulnerability: Coastal Dynamics od Dutch Dikes. *The Geographical Journal* 164 (3): 259–268.

Klein, R.J. et al. 2003. Resilience to natural hazards: how useful is this concept ? *Global Environmental Change* 5: 35–45.

Leone, F. 2008. *Caractérisation des vulnérabilités aux catastrophes « naturelles »: contribution à une évaluation géographique multirisque.* HDR, Université Paul Valéry – Montpellier III.

Lhomme, et al. 2010. Les réseaux techniques face aux inondations ou comment définir des indicateurs de performance de ces réseaux pour évaluer la resilience urbaine. *Bulletin de l'Association de Géographes Français – Géographies* 4: 487–502

Lim, B. & Spanger-Siegfried, E. (eds). 2005. *Adaptation Policy Framework for Climate Change: Developing Strategies, Policies and Measures.* UNDP and GEF, Cambridge: Cambridge University Press.

Luers, A.L. 2005. The surface of vulnerability: an analytical framework for examining environmental change. *Global Environmental Change* 15: 214–223.

Magnan, A. 2009. *La vulnérabilité des territoires littoraux au changement climatique: Mise au point conceptuelle et facteurs d'influence.* Paris: Iddri Analyses 1.

Manyena, S.B. 2006. The concept of resilience revisited. *Disasters* 30 (4): 433–450.

Maret, I. & Cadoul, Th. 2008. Résilience et reconstruction durable: que nous apprend La Nouvelle-Orléans ? *Annales de géographie* 117 (663): 104–124.

November, V. 2006. Le risque comme objet géographique. *Cahiers de géographie du Québec* 50 (141): 289–296.

Ouallet, A. 2009. Vulnérabilités et patrimonialisations dans les villes africaines: de la préservation à la marginalisation. *Cybergeo,* URL: http://www.cybergeo.eu/index22229.html

Perry, R.W. 2006. What is a disaster. In H. Rodriguez, E.L. Quarantelli, R. Dynes (eds.), *Handbook of disaster Research.* New York: Springer.

Pigeon, P. 2005. *Géographie critique des risques.* Paris: Economica.

Pimm, S.L. 1984. The complexity and stability of ecosystems. *Nature* 307 (26): 321–326.

Provitolo, D. 2002. *Risque urbain, catastrophes et villes méditerranéennes.* Thèse de doctorat en géographie, Université Nice-Sophia Antipolis, 365 p.

Provitolo, D. 2007. La vulnérabilité aux inondations méditerranéennes: une nouvelle démarche géographique. *Annales de Géographie* 1 (653): 23–40.

Provitolo, D. et al. 2009a. Vers une ontologie des risques et des catastrophes: le modèle conceptuel. In *Ontologie et dynamique des systèmes complexes, perspectives interdisciplinaires, Rencontres interdisciplinaires sur les systèmes complexes naturels et artificiels, janvier 2009.* Online: http://www.gemas.fr/dphan/rochebrune09/papiers/ProvitoloDamienne.pdf

Provitolo, D. et al. 2009b. Validation of an ontology of risk and disaster through a case study of the 1923 Great Kanto Earthquake. In C. Bertelle et al. (eds), *ICCSA Proceedings Special Sessions, 3rd International Conference on Complex Systems and Applications, 29 june-2 july 2009, Le Havre (France)*.

Provitolo, D. et al. 2011. Emergent human behaviour during a disaster: thematic versus complex systems approaches. In M.A. Aziz-Alaoui et al. (eds), *Emergent Properties for Natural and Artificial Complex Systems – European Conference on Complex System, 12–16 sept. 2011, Vienne*.

Quarantelli, E.L., et al. (eds). 2006. A heuristic approach to future disasters and crises: New, old, and in-between types. In Rodriguez, H.et al. (eds), *Handbook of Disaster Research*. New York: Springer.

Reghezza-Zitt, M. 2009. Réflexions autour de la vulnérabilité: définition d'une approche intégrée à partir du cas de la métropole francilienne. In S. Becerra, A. Peltier (eds), *Risques et environnement: recherches interdisciplinaires sur la vulnérabilité des sociétés*. Paris: L'Harmattan.

Revet, S. 2009. De la vulnérabilité aux vulnérables. In S. Becerra, A. Peltier (eds), *Risques et environnement: recherches interdisciplinaires sur la vulnérabilité des sociétés*. Paris: L'Harmattan.

Resilience Alliance, http://www.resalliance.org/1.php

Rodriguez, H.et al. (eds). 2006. *Handbook of Disaster Research*. New York: Springer.

Rufat, S. 2008. *Transition post-socialiste et vulnérabilité urbaine à Bucarest*, Thèse de doctorat, ENS Lyon.

Ruin, I. 2010. Conduite à contre-courant et crues rapides. Le conflit du quotidien et de l'exceptionnel. *Annales de géographie* 674 (4): 419–432. Paris: Armand Colin.

Smit, B. & Wandel, J. 2006. Adaptation, adaptive capacity and vulnerability. *Global Environmental Change* 16 (3): 282–292.

Thiétart, R-A. 2000. Management et complexité: concepts et théories. *Cahier 282, Centre de Recherche DMSP*.

Thouret, J.C. & D'Ercole, R. 1996. Vulnérabilité aux risques naturels en milieu urbain: effets, facteurs et réponses sociales. *Cah. Sci. hum.* 32 (2): 407–422.

Uitto J. I. 1998, The geography of disaster vulnerability in megacities: a theoretical framework, *Applied Geography*, 18 (1): 7–16.

Van der Leeuw, S E. & Ashan Leygonie, Ch. 2001. A long-term perspective on resilience in social-natural systems. *Working papers of the Santa Fe Institute*, n° 01–08–042.

Veyret, Y. 2004. *Les risques*. Paris, Breal.

Vis, M. et al. (eds.) 2001. *Living with floods. Resilience strategies for flood risk management and multiple land use in the lower Rhine River basin*. Delft: NCR.

Walker, B.H. et al. 2004. Resilience, Adaptability and Transformability in Social-ecological systems. *Ecology and Society* 9 (2).

Walker, B.H. et al. 2006. A handful of heuristics and some propositions for understanding resilience in socio-ecological systems. *Ecology and Society* 11(1).

Walker, B.H. & Salt, D. 2006. *Resilience thinking: Sustaining Ecosystems and people in a Changing World*. Washington, Covelo, London: Island Press.

Weichselgartner, J. 2001. Disaster mitigation: the concept of vulnerability revisited. *Disaster Prevention and Management* 10 (2): 85–94.

Winograd, M. 2004. Evaluations et Indicateurs de Vulnérabilité: Comment transformer les données en réponses politiques et actions d'adaptation. *Vulnerability and Adaptation Training Project*. Genève: UNITAR.

Wisner, B. 1998. Marginality and vulnerability: why the homeless of Tokyo don't count in disaster preparations. *Applied Geography* 18 (1): 25–33.

Wisner, B. 1999. There are worse things than earthquakes: hazard, vulnerability and mitigation capacity in greater Los Angeles. In J.-K. Mitchell, Crucibles of Hazard: Disasters and Megacities in Transition. Tokyo, New York, Paris: United Nations University Press.

Wisner, B. et al. 2004. At Risk: Natural Hazards, People's Vulnerability and Disasters (second ed.). New York: Routledge.

Resilience and Urban Risk Management – Serre, Barroca & Laganier (eds)
© 2013 Taylor & Francis Group, London, ISBN 978-0-415-62147-2

Using the geo-archaeological approach to explain past urban hazards

E. Fouache
University of Paris-Sorbonne, France
Senior Member of the IUF, France

ABSTRACT: The first cities emerged in the Middle East around the year 3000 BC. The geo-archaeological approach allows us to study environmental processes in an archaeological context and thus to identify the past urban hazards. There is much to be gained: these studies are fundamental to a better understanding of present-day hazards, to urban development, but also to remembering our heritage. Cities have always been susceptible to nature's risks and natural disasters, but have also—through urban development and through the close proximity of great numbers of human beings-, thrown up their own hazards.

Keywords: city, hazards, geo-archaeology, archaeology, environment, urban development, heritage

1 INTRODUCTION

If we see the end of the process of neolithisation as being around the year 6000 BC, in the Middle East first of all, then cities begin to appear after a long maturation period around 3000 BC. They sprung up on the banks of big rivers in Mesopotamia, in the areas that are now Iraq, Iran and Syria; and in the valleys of the River Nile, the River Jordan, the River Indus, the River Ganges and the Yellow River. The date of their emergence in Africa, Meso-America and South America is still disputed, but is definitely more recent. What distinguishes cities from villages or clusters of little farming towns is the concentration in a single area of economic, political, social and religious powers, which was reflected in a high relative concentration of non-farming populations and in a monumental administration often enclosed within a surrounding wall.

Geo-archaeology is an interdisciplinary approach that we define as being founded upon the use of methods and techniques drawn from geosciences, archaeology and geography, to recreate—in a multi-scalar, diachronic archaeological perspective-, palaeoenvironments and landscape processes, in connection with human inhabitation (Fouache, 2010). The geo-archaeological approach allows us to comprehend, in an urban studies context, the evolution of environmental processes over the course of time, and the parallel evolution of the issues, the vagaries and thus the hazards of a city. Large modern-day developments—like the construction of an underground car-park in Lyon (Quai Saint Antoine), or the boring of tunnels in Istanbul (in the Yenikapi neighbourhood), going under the Golden Horn and the Bosphorus to link old Stamboul to the Asian side of the city-, have resulted in a proliferation, where ancient cities once stood, of conservation-oriented archaeological excavations associated with environmental studies; in this instance the identification of the bed of the River Saône in the early Iron Age (http://lugdunumactu), and the excavation of a 10th–11thC AD Byzantine port (Degremont, 2009). This is a major opportunity to bring together urban development, archaeology and history; and to encourage understanding of environmental processes amongst urban populations by integrating the notion of heritage into this understanding. This is of importance to the whole world, and especially to the world's major cities.

Explaining the hazards of ancient cities can also serve as a great warning to the current generation, and promotes prevention policies rooted in a real understanding of the interaction between natural processes and processes brought about by human societies. We mustn't, however, fall into any irrational fear of the risks of our environment. Between the Bronze Age in the Middle East, the Iron Age in Western Europe and the end of the modern era—to stick to a more archaeological than historical approach-, examples of urban civilizations that have been destroyed solely by environmental disasters are rare.

2 CITIES AND CATACYSMS

The greatest risk for cities of the past was, of course, as today, a cataclysm; be it the direct effects of a natural catastrophe or the effects of society. The most notable kinds of such cataclysms have to be volcanic eruptions, earthquakes and tsunamis (Fouache, 2006). Numerous cities were rubbed off the map in this way: Pompeii, Herculaneum and Stabiae in Campania when Mount Vesuvius erupted in AD 79; Akrotiri in Santorini between 1635 and 1628 BC; and Helike taken by the Corinthian Gulf in 373 BC. The Lisbon earthquake of 1755, followed by a tsunami and the destruction of the city by the ensuing fire, is a good example of how urban development from human beings can heighten the consequences of a catastrophe. The Fukushima catastrophe is another in this long line. However, a city being erased from the map is somewhat exceptional; in the past as in the present, cities normally rise from the ashes in the same place, providing that the political and social systems dispose of sufficient resources to do so, and that the city is needed by the socio-economical, political and religious systems. It is due to these socio-economical, political and/or religious imperatives that the city is rebuilt in the same place, whether because of the site or because of the image of the city. A city's exposure to 'natural' hazards is not, however, limited to cataclysms.

3 NON-CATACLYSMICS NATURAL HAZARDS

The frequent location of cities on the banks of rivers (Bravard and Magny, 2002) or by the sea puts them at the mercy of drastic rises in water levels, floods and changing shorelines (Morhange et al, 2007). The original site of the city is often sheltered from these dangers, on a headland or hill; but as soon as the city has been established, a lower city, suburbs and a port, which are exposed, all come into existence; then, very soon after, urban growth spreads into at-risk areas, the original site being too small. An example of this is the town of Sommières in the Gard department (France), built at the foot of the hill on which the castle stands, right on the River Vidourle flood plain despite the high frequency of floods originating in the Cevennes rains. The adoption of collective risk-management measures can be a factor in social acceptance of the hazard. Thus the Ancient Egyptians viewed the floods of the Nile as a nourishment of the land, and the Venetians very quickly learnt to live with the phenomenon of Acqua Alta.

Indeed, the site chosen can itself be the cause of a city's wealth. The ancient city of Mari (Margueron, 2004) on the Middle Euphrates in modern-day Syria was closely linked to the river, its irrigated farmland and its waterway. Throughout the third and second centuries BC, these advantages enabled costly development that the Sumerian civilisation in Uruk had the human and financial resources to support. The surrounding wall of the Mari palace can thus just as easily be viewed as a fortification as it can be a dyke. The same is true for Babylon on the River Tigris: from the late 17th to the 11th century BC, Babylonian kings were incessantly obliged to raise their ramparts, nearly 20 m in total, to fight the intense aggradation of the river. In the end it wasn't the river's water levels that did for Babylon, but Median invasion.

What geo-archaeological studies also teach us, as well as the studies of paleo-climatologists, is that at the level of the Holocene (Mayewski et al., 2004)—the last 10,000 years-, environmental processes have varied: the seasonal distribution of weather types, temperatures (by a yearly average of roughly two degrees), rainfall (by irregular amounts) (Birck et al., 2005); the consequences of all of this are heightened all the more in remote areas of the inhabited world. As a result, the morphogenesis and the hydrological rhythms of rivers and thus the hydro-morphological hazards have evolved (Arnaud-Fassetta, 2000, 2008), between calm periods such as the Medieval Warm Period, and other periods witnessing far more exceptional occurrences, such as the Little Ice Age. In the same time period, the changes societies have brought to the exploitation of drainage basins have interacted with these natural processes, sometimes worsening environmental disasters, sometimes balancing them out (Diamond, 2006). This is how—due to erosion linked to the agricultural exploitation of its hinterlands and the progradation of the Küçük Menderes Delta-, the city of Ephesus ended up ceasing to be a port (Kraft *et al.*, 2007). However, cities do not merely suffer their environment, they affect it themselves.

4 THE IMPACT OF CITIES ON THEIR ENVIRONMENT

From the very beginning of its life, a city has a considerable impact on its environment. Huge numbers of inhabitants—Xi'an, for example, in the Chinese province of Shaanxi, is believed to have been the home of over a million people in 1000 BC-, have always been conducive to epidemics (Hays, 2005), pollution (Botsos *et al.*, 2003), and a heightened consumption of energy and natural resources. The supposed plague—more probably a typhus epidemic-, that hit Athens from 430 to 427 BC is still remembered today. Such paleopollution appears in geo-archaeology in the study of intra-site sediments, or that of sedimentary archives in ancient port basins and lake and river sediments found downstream from cities. Thus geo-archaeologists use levels of lead, scoria and heavy metals to trace and mark out the past, whilst the study of skeletons in necropolis offers a glimpse of the health of an ancient population and the impact of chronic illnesses and epidemics. The most finely developed ancient urban societies took measures against some of these risks, as far as the knowledge of their era allowed them to. One needs look no further than supplies of clean drinking water; fountains; urban water tanks; qanat networks (Briant, 2001)- which became commonplace across the Iranian plateau and beyond in the first millennium BC-, or Roman aqueducts (Bonnin, 1985); waste disposal and wastewater disposal; the establishment of cemeteries outside of the city; legislation aiming to curb the risk of fire; paraseismic construction practices; the building of dykes and levees; or the confinement of polluting activities or industries, for example tanneries situated in their own specific neighbourhoods. We must be careful, however, not to apply modern standards of safety and responsibility to ancient urban societies.

5 THE NEW RESPONSIBILITY OF CONTEMPORARY URBAN SOCIETIES

In an epidemiological as well as an environmental sense, there is most definitely a difference in size between ancient and modern urban societies. Our civilisation—with its knowledge of tectonic plate theory and its vast advances in geosciences and biology-, is the first to have gained a scientific understanding of the genesis of volcanic eruptions, earthquakes, tsunamis and landslides, as well as of the environmental disasters or health crises of the past, and the first to have conceived of forecasting such phenomena. Considering current scientific and technological expertise, preventing major and natural hazards should be a top priority for cities all over the world, based on a five-point plan: studying dangers, knowing what is at stake, defining risks, adopting urban planning regulations, and educating city-dwellers on the states of crisis specific to each urban context. Be it due to a lack of specialists, a lack of resources, an absence of political will, corruption, or misunderstood financial interests, such a plan is still a rare exception in the world today.

6 WHAT CAUSES A CITY TO DISAPPEAR?

Seeing as cities are each built on areas with their own topography, with site constraints and hazards linked to these dangers, they are in essence social, economical, religious and political products. By way of proof, all cities, throughout history, have been created either by a myth or by a decree. In the Middle East, the first cities to be recognised and partially explored were, obviously, new cities (Margueron, 2004), clearly built by political will. Later on, foundations were in fact often found to be hiding refoundations, and the desire of authority to forever leave its mark on history. The city of Kar-Tukulti-Ninurta (Eickhoff, 2005), is emblematic in this sense: situated on the right bank of the River Tigris, where modern-day Iraq stands, it was founded in such a way by King Tukulti Ninurta I, who ruled from 1244 to 1208 BC. This veritable new city boasted a design based on quadrilaterals, long orchard paths and baked-brick ramparts. The king wanted it to be a new capital, but he died before it was finished, and the city soon crumbled.

What archaeology teaches us is that cities disappear with the civilisations that founded them, for reasons often far more social, political or religious than environmental. The aridification of the global climate (Kuzuçuoglou and Marro, 2007; Fouache et al., 2009) that followed the Holocene Climate Optimum is often put forward as the reason behind the great cultural crises that occurred in the Near East, such as the fall of the Akkad Empire (Weiss et al., 1993) in the Arabo-Persian Gulf, or the abandonment of Harappan cities in the Indus Valley. This aridification was most probably caused either by thousand—or hundred-year fluctuations, by some sudden turn of events, or by a progressive evolution of the climate towards dryness, connected with the waning of the Indian monsoons (Lézine et al., 2007). Pollen and speleothem analyses in the region do show that this waning of the Indian monsoons from 4700 to 4200 BP (Ivory and Lézine), was very real, but an actual link with the collapse of late Bronze Age civilisations is not certain.

When we discuss the collapse of Bronze Age civilisations in the Near and Middle East, we thus must take care to distinguish vast urban centres—economic, political and cultural hubs, very dependant on external flows, and, as we have intimated, very quickly abandoned at the end of the third millennium-, and little urban and rural sites like those, for example, of the Sabzevar region (Fouache et al., 2010) in Iran, which have been occupied consistently throughout their existence. One must also consider the duration of aridification, which is by no means a quick and brutal process, but rather a slow evolution over the course of 600 years. To attribute the collapse of these Bronze Age civilisations to the only climactic factor in question seems, in the light of current archaeological knowledge, a gross simplification.

If we move to Central Asia, and protohistoric Central Asia, recent archaeo-environmental research (Cattani, 2005; Francfort, 2005; Francfort, 2009; Francfort and Tremblay, 2010; Luneau, 2010) seems to be showing us that the pinnacle of Oxus civilisation came at the end of this phase of aridification, proven by environmental studies (Cremaschi, 1998), notably the advance of the Kara-Kum dunes and the parallel southerly advance of steppe peoples (Cattani, 2005); its collapse, meanwhile, most likely came with the beginning of a new humid phase.

7 CONCLUSION

We are fortunate to be able to use conservation-oriented archaeological digs to carry out, within a dense urban tissue, geo-archaeological studies of environmental processes. This affords us a better understanding of initial site constraints, and allows us to piece together the dynamic evolution of environmental constraints which interacted with processes begotten by human development. This piecing together in turn affords us a better understanding of the perennial or random nature of hazards, lets us know when they will recur, and allows us to form more effective prevention policies. This environmental history can also form part of a city's heritage, a part that can be exhibited for all to see and—by placing a city's current situation in a dual history, that of the environment and that of human societies-, can be used for educational purposes in explaining the nature of hazards and the evolution of issues and risks. We must not, however, focus exclusively on environmental risks to our cities. Preventing major hazards, reducing pollutants and managing waste, optimising the management of water resources, and harnessing the growth of megalopoles are all major issues; but—due to the huge numbers of inhabitants and their increasing concentration-, the greatest dangers to the long lives of our cities are actually, now as they always have been, social and political.

ACKNOWLEDGEMENTS

I am most particularly indebted in the preparation of this article to Annie Caubet, the honorary General Curator of the Louvre Museum, for her remarks and suggestions on Middle Eastern cities and Joe Cunningham for translation.

REFERENCES

Arnaud-Fassetta, G. 2000. 4000 ans d'histoire hydrologique dans le delta du Rhône. De l'Âge du bronze au siècle du nucléaire. Grafigéo, 11, collection mémoires et documents de l'UMR PRODIG, Paris, 229 p.

Arnaud-Fassetta, G. 2008. La Géoarchéologie Fluviale. Concepts, attendus et méthodes d'étude rétrospective appliquées à la caractérisation du risque hydrologique en domaine méditerranéen. Écho-Géo, 4, p. 2–11.

Borsos, E. Makra, L. Beczi, R. Vitanyi, B. Szentpéteri, M. 2003. Anthropohenic air pollution in the ancient times. Acta Climatologica et Chrorologica. Tom. 36–37, 5–15.

Bravard, J.-P. Magny, M. 2002. Les fleuves ont une histoire. Paléo-environnements des rivières et des lacs français depuis 15000 ans. 312 p.

Birck, J. Battarbee, R. Macay, A. Oldfiel, F. 2005. Global Change in the Holocene. Holder Arnold. 480 p.

Bonnin, J. 1985. L'eau dans l'Antiquité. Eyrolles, 488 p.

Briant, P. (Ed.) 2001. Irrigations et drainages dans l'Antiquité, Qanats et canalisations souterraines en Iran, en Egypte et en Grèce. Persika 2, Paris, 190 p.

Cattani, M. 2005. Margiana at the end of Bronze Age and beginning of Iron Age. In M.F. Kosarev, P.M. Kozhin, and N.A. Dubova (eds.). *Uistokov civilizacii. Sbornik statej k 75-letiju Viktora Ivanovicha Sarianidi,* Moscou, Kollektiv avtorov, p. 303–315.

Cremaschi, M. 1998. Palaeohydrography and Middle Holocene Desertification in the Northern Fringe of the Murghab Delta. In A. Gubaev, G.A. Koshelenko, et M. Tosi (eds.). *The Archaeological Map of the Murghab Delta. Preliminary Reports 1990–95,* Reports and Memoirs, vol. Series Minor Volume III, Rome, Istituto Italiano per l'Africa e l'Oriente. Centro Scavi e Ricerche Archeologiche, p. 15–25.

Degremont, C. 2009. Istanbul: le port byzantin de Yenikapi. Archeologia, n° 469, p. 16–25.

Fouache, E. 2006. 10000 d'évolution des paysages en Adriatique et en Méditerranée Orientale. Travaux de la Maison de l'Orient Méditerranéen (TOM), Volume 45. Lyon; Paris: diff. De Boccard. 225 p.

Eickhoff, T. 1985. Kãr-Tukulti-Ninurta, Eine mittelassyrische Kult und Rezidenzstadt, ADOG 21, Berlin.

Fouache, E. Lezine, A.M. Adle, S. Buchsenschutz, O. 2009. Le passé des villes pour comprendre leur futur. In *Villes et géologie urbaine,* Géosciences, 10. p. 54–61.

Fouache, E. 2010. L'approche Géoarchéologique. In H. Alarashi, M.-L. Chambrade, S. Gondet, A. Jouvenel, C. Sauvage et H. Tronchère (eds.). *Regards croisés sur l'étude archéologique des paysages anciens. Nouvelles recherches dans le Bassin méditerranéen, en Asie Centrale et au Proche et au Moyen-Orient?.* Lyon, Maison de l'Orient et de la Méditerranée Jean Pouilloux, 2010.- 256 p.: nombreuses ill.; 30 cm.- (Travaux de la Maison de l'Orient; 56). p. 17–30.

Fouache, E. Cosandey, C. Francfort, H.P. Bendezu-Sarmiento, J. Vahdati, A.A. Lhuillier, J. 2010. The Horst of Sabzevar and regional water ressources from the Bronze Age to the present day (Northeastern Iran). Geodinamica Acta. 23/5–6, p. 287–294.

Francfort, H.P. 2005. La civilisation de l'Oxus et les Indo-Iraniens et Indo-Aryens. In G. Fussman, J. Kellens, H.-P. Francfort, and X. Tremblay (eds.). *Aryas, Aryens et Iraniens en Asie Centrale.* Collège de France. Publications de l'Institut de Civilisation Indienne, vol. 72, Paris, Diffusion de Boccard, p. 253–328.

Francfort, H.P. 2009. L'âge du bronze en Asie centrale. La civilisation de l'Oxus. Anthropology of the Middle East, 4 (1), p. 91–111.

Francfort, H.P. & Tremblay, X. 2010. Marhai et la Civilisation de L'Oxus. Iranica Antiqua, XLV, p. 51–224.

Hays, J.N. 2005. Epidemics and pandemics: their impacts on human history. 513 p.

Ivory, S.I. & Lezine, A.M. 2009. Climate and environmental change at the end of the Holocene Humid Period: a pollen record off Pakistan, CR Geoscience.

Kuzuçuoglou, C. & Marro, C. (Eds.). 2007. Sociétés humaines et changements climatiques à la fin du troisième millénaire: une crise a t'elle eu lieu en Haute Mésopotamie?. Actes du colloque de Lyon, 5–8 décembre 2005. Varia Anadolica 19, 590 p.

Luneau, E. 2010. L'Age du Bronze Final en Asie Centrale Méridionale (1750 = 1500//1450 avant n.e.): la fin de la civilisation de l'Oxus. Thèse nouveau régime. Université de Paris 1. 612 p

Kraft, J.C. Brückner, H. Kayan, I. Engelmann, H. 2007. The geographies of Ancient Ephesus and the Artemision in Anatolia. Geoarchaeology, 22, N°1, p. 121–149.

Lézine, A.M. Tiercelin, J.J. Robert, C. Saliège, J.F. Cleuziou, S. Inizan, M.L. Braemer, F. 2007. Centenial to millennial-scale variability of the indian monsoon during the early Holocene from sediment, pollen, and isotope record from the desert of Yemen. Palaeogeogr.Palaeoclimatol.Palaeoecol. 243. p. 235–249.

Margueron, J.C. 2004. Mari: métropole de l'Euphrate au IIe et au début du IIe millénaire av. J.-C., ERC, 575 p.

Mayewski, P. Rohling, E. Stager, J. Karlen, W. Maasch, K. Meeker, L. Meyerson, E. Gasse, F. Van Kreveld, S. Holmgren, K. Lee-Thorp, J. Rosqvist, G. Rack, F. Staubwasser, M. Schneider, R. Steig, E. 2004. Holocene climate variability. Quaternary Research 62, p. 243–255.

Morhange, C. Marriner, N. Sabatier, F. Vella, C. 2007. Risques littoraux en Méditerranée. Méditerranée, 108, 149 p.

Veyret, Y. 2009. Du risque à la gestion des villes: la ville durable. In Villes et géologie urbaine, Géosciences, 10. p. 94–101.

Weiss, H. Courty, M.A. Wetterstrom, F. Guichard, F. Senior, L. Meadow, R. Curnow, A. 1993. The Genesis and Collapse of the Third Millenium North Mesopotamia Civilization, Science 261, p. 995–1004.

Internet: http://lugdunumactu.wordpress.com/2010:12/11/fouilles-archéologiques-du-quai-st-antoine-lyon/

Resilience and Urban Risk Management – Serre, Barroca & Laganier (eds)
© 2013 Taylor & Francis Group, London, ISBN 978-0-415-62147-2

Evolution of natural hazard assessment and response methods

R. Ashley, J. Blanksby & A. Saul
Pennine Water Group, University of Sheffield, UK

B. Gersonius
Flood Resilience Group, UNESCO IHE, Delft, The Netherlands

ABSTRACT: The paper considers the development of assessment of natural hazards and response methods with special reference to flooding, which may be considered as a socio economic hazard rather than a natural one. Set within the context of climate change, the paper considers how current flood risk is assessed and the additional uncertainties that occur when assessing future flood risk The effect of uncertainty on flood risk managemenet measures, including insurance is considered.

1 INTRODUCTION

What are natural hazards? We all think of floods, droughts, earthquakes, hurricanes, volcanic eruptions, extraterrestrial objects, forest fires etc. Whilst natural hazards impact on the planet and on ecosystems, the term has little meaning for us unless taken in an anthropocentric context. As humans are also 'natural', as emphasised by the 'humans-in-nature' concept (Berkes & Folke, 1998) means that the hazards created by humans may also be considered to be in the categories of interest. In pre-history, natural phenomena, such as those itemised above, were the primary cause of hazards, risks and change to human existence; with ecosystems being as much a threat as a source of sustenance, as getting killed by wild animals was a constant worry. Assessment and response to these threats in those days comprised selecting living locations, migration and the evolution of weapons and enclosed living spaces. Interestingly, lifestyle was just as important as it is today in managing exposure, vulnerability and ensuring resilience to hazards. The difference then was there were fewer people on the planet and ownership of goods and services (inventories) was much less significant.

Nowadays there seem to be an increasing number and scale of threats from natural hazards at both acute (e.g. volcanic eruptions, floods) and chronic (e.g. climate change induced effects) timescales as global system complexity is seemingly increasing (Gunderson, 2010). 'Catastrophes' are also cited as growing worldwide: *"Of the 25 most costly insured catastrophes in the past 40 years, two-thirds have occurred since 2001"* (World Economic Forum, 2011) and in the USA (Michel-Kerjan & Kunreuther, 2011) state that the cumulative expected exposure of the U.S. government to catastrophes[1] over the next 75 years could reach $7 trillion. The term 'catastrophe' is used extensively as meaning a major impacting event but is defined in different ways in different countries and different contexts: *'Catastrophe insurance is different from other types of insurance in that it is difficult to estimate the total potential cost of an insured loss and a catastrophic event results in an extremely large number of claims being filed at the same time'* (Investopedia, 2011), suggests that a catastrophe has a huge impact and affects a lot of recipients.

Even where 'catastrophes' are clearly 'natural' and have and continue to (often violently) shape landscape, populations often try to cling on. An example of one of the most catastrophe

---

1. Of course in the US context this also includes terrorism.

prone places on earth, the Kuril islands, is subject to multiple and frequent 'natural' cata-strophic impacts (just about all of those in the introduction above). Notwithstanding the hos-tility of the islands, both Russia and Japan lay claim to the land and 16,800 inhabitants (2003) (Ganzey et al, 2011). There are even 'great catastrophes: '*In all, there were five catastrophes last year assignable to the top category of "great natural catastrophes" based on the definition criteria of the United Nations: the earthquakes in Haiti (12 January), Chile (27 February) and central China (13 April), the heatwave in Russia (July to September), and the floods in Pakistan (also July to September). These accounted for the major share of fatalities in 2010 (around 295,000) and just under half the overall losses caused by natural catastrophes.'* (Munich Re, 2011).

Disasters, catastrophes, hazards and hazard assessment can be viewed at a variety of scales both spatial and temporal (e.g. Birkmann & von Teichman, 2010). Some hazards are clearly global in potential effects and impacts, such as climate change, whereas others are localised, such as local flooding or small magnitude earthquakes. On a temporal scale some potential hazards are seen in a perspective that is way beyond human existence such as the potential end of the Holocene period that has given us very stable global weather patterns for some 10,000 years (Foley et al, 2003) and could come to an end virtually without warning in what would be '*the* great catastrophe'. For today's humans, it is the acute and localised hazards that are usually of most concern, and may be the hazards for the next generation or so. This is why effecting behavioural change to lifestyles that are less likely to provoke 'natural' hazards in the future is so difficult; essentially we are self-centred and cannot see far enough to worry about longer-term chronic hazards that are exacerbated by our behaviour today but do not impact upon our lifestyles, only of those of future generations (e.g. Barr et al, 2011).

There is an increasing awareness of hazards and threats, especially to the recent consump-tive and acquisitive lifestyles in the Western world, now impacted by the economic 'hazard' (is this a 'natural hazard'?). Many 'natural hazards' are in fact simply consequences of lifestyles and the way in which human society is structured—in both cause (e.g. global warming; lack of morality) and exposure (e.g. living in the wrong place; taking too many risks). Unfortu-nately for many communities, lifestyle, societal structures and associated behaviours are often not a choice, especially for the poorest (Schad et al, 2011). Hence hazard impacts are mainly a function of human vulnerability which is now understood to be a product of specific spatial, socio-economic-demographic, cultural and institutional contexts (Kuhlicke et al, 2011). In its report "Global Risks 2011" the World Economic Forum identified that the issue with the highest level of risk was climate change, an environmental problem. The next three highest risks were perceived as being socio economic and political with the fifth highest ranked as storms and cyclones whereas flooding was ranked ninth. Both storms and cyclones and flood-ing are classed as environmental. In parallel an assessment of risk interconnectedness is also illuminating in that only climate change out of the environmental sources of risk, only climate change is included in the top ten, coming in at number eight. and many of the 'natural' haz-ards are regarded as being peripheral. Furthermore the report identifies that out of three key nexus' of interconnected risks, only the food-water-energy nexus includes natural systems.

A key distinction needs to be made between the concept of risk and that of uncertainty in the context of this paper. Risk applies to a situation where the probability and impact of an adverse event can be inferred from past behaviour. Risk would then apply to a situation involving the likelihood of damage. Because the expected costs of such risks are quantifiable using statistical and other methods, resources can be set aside to insure against or prepare/recover from the potential damage. Uncertainty, however, is here considered a condition in which the probability and scale of adverse events cannot be inferred from past information, i.e. in the case of climate change as an example, the perceived threats and impacts are de facto extrapolated from the past and used to define what may be anticipated. This means that there are uncertainties identified from our knowledge of the past and in addition uncertainties in the analytical process, which reduce the certainty of the predictions of probability. Hence it becomes more appropriate to use the qualitative term likelihood rather than the quantitative term probability. The uncertainty can arise in a category of "known unknowns". Awareness can exist concerning the possibility of a catastrophic event, but with few clues regarding whether such a disaster could set in motion irreversible processes beyond certain thresholds.

This form of uncertainty is incorporated, for instance, into many future climate change scenarios. Yet, even where risks can be quantified, the impact of interactions among these risks could still be uncertain (United Nations, 2011).

In this paper the evolution in understanding of natural hazards is briefly reviewed, considering mainly flooding risks. Hazard assessment and potential responses are also considered, especially the recent developments in understanding of consensual based non-structural measures delivered in partnership with vulnerable communities.

2 UNDERSTANDING HAZARDS AND RISKS

Understanding and assessing natural hazards has now evolved from considering nature as 'external' to humanity and often a threat (an act of God) (e.g. Vaught, 2009) to one of matching probability of occurrence, exposure, vulnerability and impacts to risk acceptability and human tolerance and interaction with natural systems (Oberholzner, 2011; Pantti & Wahl-Jorgensen, 2011). Although there are definition issues about terminology in the way this is seen (e.g. Fuchs et al, 2011). Of course, not all of humanity has abandoned the 'Act of God' view, not least many in the insurance sector, where an act of God is still included in disclaimers in many policies. In a recent flood awareness survey carried out with 63 engineering students at the University of Sheffield and 631 professionals working for the City of Bradford Metropolitan District Council, only 2 responders held God responsible and only 2 held nature responsible. All the rest held that human beings were the source of flooding for one reason or another.

Thus the understanding of the importance of considering social-technical systems, rather than 'natural/environmental' systems or 'technical' systems has grown (Fuchs et al, 2011). This is further explained in the section on assessing flood risk and responding. In order to make a meaningful contribution to the discussion on hazards in this paper, floods and flooding are now focused on.

There are many definitions of hazards and it is easy to cause confusion through inappropriate use of language so it is important to be clear about what we mean. For instance, flooding may be widely defined; the EU Floods Directive (EU, 2007) defines "flood" as the *temporary covering by water of land not normally covered by water*, and "flood risk" as the *combination of the probability of a flood event and of the potential adverse consequences*. Clearly the former of these can relate to any form of inundation, affecting humans or not, whereas the latter is clearly anthropocentric. Whether or not the flood event is a catastrophe also depends on individual impacts and how insurers define the term in the context of a given event. The definition of flooding above is only useful provided everyone concerned with floods in Europe has read and understood the meanings in the flood directive. However, widespread ignorance of such standard definitions enables people (and e.g. insurers) to use the term 'flood' in diverse ways for their own aims. In order to avoid this it may be useful to use a different term altogether (such as high water) for the temporary and often natural and beneficial covering of land by water, and preserve the term flood for the interaction between humans and water, which is largely influenced by human choice. However, since the publication of the Flood Directive in 2007, the International Standards Organisation has published its principles and guidelines (ISO 2009a) and vocabulary (ISO 2009b) and these have been adopted by many countries in their national Norms and Standards. This means that for a minimum of effort it will be possible to develop consistency of language and process across national and disciplinary boundaries for the management of risk.

As well as considering the balance between natural and human input into the creation of a catastrophe, in order to develop appropriate responses, it is also helpful to consider the distribution and nature of the impacts of a catastrophe or a series of catastrophes associated with different hazards. This is illustrated in Table 1.

In Table 1 the assessment of the severity of the risk and the potential to avoid it (Low, Medium, High) is subjective, but the message is that a hazard can have different consequences for people than for buildings and property. Drought (and famine are a good example of this where the impact on people is potentially immense whereas that on property is by and large minimal.

Table 1. Water related natural hazards and natural and 'non-natural' risks in Europe.

Hazards	Risk to people	Avoidability of risk	Risk to buildings and property	Avoidability of risk
Storm	H	L	H	L
Driver is influenced by human activity (emissions and climate change)	There is no hiding from a storm. Once it starts everyone and everything within its vicinity is potentially vulnerable. Building codes can help. Largely a natural risk.			
Earthquake	H	L	H	L
Driver not influenced by human activity	There is no hiding from an earthquake. Everyone and everything within its vicinity is potentially vulnerable. Building codes can help. Largely a natural risk.			
Drought	H	M	L (Agriculture has high risk)	H
Driver is influenced by human activity	It affects everyone in an area. Needs reservoir capacity. Largely a natural risk in terms of driver (too little rain) but in terms of impact is anthropocentric.			
Heat wave	H	L	L	H
Driver is influenced to some extent by human activity	Affects everyone in an area, heat island compounds this, but reduces the affects of cold. Largely a natural risk.			
Cold	H	M	L–M	H
Driver is influenced to some extent by human activity	Affects everyone in an area, heat island compounds this, but reduces the affects of cold. Largely a natural risk.			
Flood	L–M	H	H	H
Driver is influenced by human activity	High water is natural, but flooding is an entirely human contrivance/hazard.			

The distribution of risk is also important to consider as this influences the attitude of a society to how to manage the risk and to those affected by the risk. A hazard that has widespread consequences is more likely to develop cohesive responses at regional, national and transnational scales whereas one that has only limited impact is unlikely to develop such cohesive responses. In the UK, some 5 million people are claimed to live or work in places that are at risk of flooding under current climatic conditions (EA, 2011; NAO, 2011), but this represents less than 10% of the population, so it is understandable that flooding is not given a high priority by individuals. In fact *every single property and person is at risk* in the UK as rain can and does fall over the entire country (Ashley et al, 2009). The latter is not reported due to the Institutional and governance structure surrounding flood risk management in the UK (Newman et al, 2011).

In contrast to flooding, the widespread distribution of earthquakes in Southern Europe is such that in time, whole countries are affected and so social cohesion is much more likely to play a part in any response (Figure 1).

Considerable efforts are now going into understanding and coping with hazards from whatever source and there is a wealth of standards (e.g. ISO 31000) dealing with living with risk. Guidance documents point to mistakes of the past (e.g. Kropp et al, 2011).

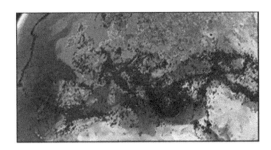

Figure 1. Distribution of M≥ 3.0 Earthquakes in Europe in the Last 2000 years (Source: AIR, Paolo Bazzurro, 2011 and presented by Eugene N. Gurenko, Ph.D., CPCU, are at EU Prevention and Insurance of Natural Catastrophes, Brussels, 18/10/2011).

Nowadays we have a new category of natural hazard triggered events (often termed disasters) referred to as *Natech* disasters that occur when an industrial accident is triggered by a natural (or man-made) event. For example, the 2008 Wenchuan earthquake in China impacted on a heavily industrialised area, causing, amongst other things the release of toxic, flammable and explosive materials which impacted the population and the environment (Krausmann et al, 2011). The latter report that *Natech* events are increasing due to an increase in hazardous natural events and also growing industrial development in many countries, some of whom are not well prepared to cope with hazards. They report that floods, earthquakes and lightning strikes increasingly cause structural damage and also the release of hazardous materials into the atmosphere, to land or aquatic environment, with the main impacts from flooding being tank rupture (in 74% of 272 events). Vulnerability of a specific equipment type and its failure modes were found to depend on the characteristics of the natural-hazard. Recommendations are made for managing the 3 types of risk in the paper.

There is also a new type of known natural risk, that due to human activity. Considerable efforts have recently been directed towards categorising and quantifying man's impacts on environmental and ecological systems e.g. (UN, 2011):

− The cumulative effects of the degradation of the Earth's natural environment has increased the scale of the sustainable development challenge enormously. Provisioning for human life using the current technology is expected to be increasingly infeasible as population continues to increase and the harmful impacts of human production and consumption multiply.
− Business as usual is not an option. An attempt to overcome world poverty through income growth generated by existing "brown technologies" would exceed the limits of environmental sustainability.

Awareness of human impacts has recently resulted in the generation of a number of global reports considering impacts, known as the Millennium Ecosystem Assessment (Reid et al, 2005) which linked human well-being to changes in the services that ecosystems provide when the ecosystems are themselves impacted (mainly by humans). This has produced a new way of looking at environmental and ecological systems.

In summary, it is apparent that the history of our understanding of natural hazards has progressed as illustrated in Figure 2, from an externally imposed and often righteous punishment to one in which we now understand that often the greatest natural hazard is man. Ironically, an adaptation response by early humans was to live a nomadic lifestyle; with the coming of floating houses and cities in the Netherlands, a peripatetic lifestyle, moving moorings from time to time, may become more common in the future (essentially returning to the nomadic life).

3 ASSESSING FLOOD RISK AND RESPONDING

"Altogether a total of 950 natural catastrophes were recorded last year, nine-tenths of which were weather-related events like storms and floods. This total makes 2010 the year with the

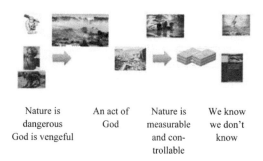

Nature is   An act of   Nature is   We know
dangerous   God   measurable   we don't
God is vengeful   and con-   know
                                   trollable

Figure 2.   Natural hazards and understanding thereof.

second-highest number of natural catastrophes since 1980, markedly exceeding the annual average for the last ten years (785 events per year). The overall losses amounted to around US$ 130bn, of which approximately US$ 37bn was insured" (Munich Re, 2011). In insurance terms two features typically characterize catastrophic risks: 'fat-tails'; and spatially correlated risks (Kousky, 2011). Many natural catastrophes have been shown to be "fat-tailed" indicating that the probability of an event declines gradually relative to its severity. Therefore the most extreme event of this type observed to date could be several times greater than the second most extreme event, which could be several times greater than the third. For most catastrophic risks losses are correlated in space, that is, a large number of receptors in close proximity are affected simultaneously. As the size of an event increases, the number of receptors affected also increases. This is very different from a risk such as theft, where the victimization of one person does not mean that a neighbour is also victimized. Furthermore, receptor types in close geographic proximity are often similar further correlating damages in the event of a flood.

Schmidt et al (2011) provide a generic framework for quantitatively assessing natural hazard risks from whatever source these originate from.

Responding to hazards by reducing vulnerability, exposure or impacts necessitates a range of measures being employed. These comprise both hard and soft measures, with new or modified infrastructure provision in the former and a wide range of non-structural measures in the latter. In the flood risk management domain, for example, these non-structural measures include changes in acceptability of living differently, land use planning and risk spreading through insurance (e.g. Dawson et al, 2011). Mitigation of natural hazards depends on the type of hazard, as ways to mitigate the occurrence and effects of volcanic eruptions (an acute hazard) are not clear other than in trivial instances, whereas the link between climate change (a chronic hazard) and human activity is now generally accepted, and mitigation by reducing greenhouse gas emissions is recommended extensively. Consequent impacts by the changing climate are then expected to be reduced (e.g. Evans et al, 2004). Balancing mitigation and adaptation to hazards is a complex challenge and rules for this have still to be properly developed (e.g. Rahman et al, 2008).

Given that hazard impacts are mainly a function of vulnerabilities and that these vulnerabilities may be addressed best within a socio-technical framework, responses are therefore combinations of hard infrastructure and other 'softer' measures, so called non-structural. Dawson et al (2011) list potential flood risk management measures when considering the place of non-structural responses to protect London, as summarised in Table 2.

In the Table the two main non-structural measures are land use planning and insurance. Each of these fits into an economic perspective on flood risk management. The former now also ties-in with the ecosystem services approaches starting to gain credence (e.g. Everard, 2011). In this paper we consider only the insurance response to living with hazards. Insurance is very reliant on the individual, organisation and national culture as to the part it may play in helping cope with natural hazards.

In many places we have moved from a top-down approach based on experts to one with much more community engagement (Newman et al, 2011) in a process of flood risk

Table 2. Summary of flood risk management measures (adapted from Dawson et al, 2011).

Intervention	Effects	Modification to risk
Climate change mitigation	May reduce degree of climate change	Changes probability of extremes globally
River and coastal engineering measures	Floods contained in the river or kept out of vulnerable area	Reduces probability of flooding in vicinity of measures. May increase it elsewhere.
Rural runoff reduction and storage	Reduces amounts of runoff and hence peak flood flow and volume	Reduces probability of flooding downstream
Flood incident management	Forecasting systems provide information to respond to allowing exposure of receptors to be reduced, possibly with temporary protection and mainly with removal from harm	Reduces frequency at which individuals personally experience flooding (they are got out of the way) and can help reduce impacts
Flood proofing	Reduces damage	Reduces impacts
Land use planning	Controlled exposure by ensuring buildings are located safely	Reduce impacts
Building codes	Reduced damage by building appropriately	Reduce impacts
Risk spreading (insurance)	Redistributes costs/impacts across population and over time	Can aid recovery and also signal need to build/live differently
Health and social measures	Reduced impacts	Usually translated into reducing economic impacts on society as a whole
Urban runoff reduction and storage	Reduces probability of flooding	Lowers flood probability

management. In England now there is an expectation that beneficiaries of local flood alleviation will contribute towards this directly, with typically 30% of the costs.

Despite exhortations for communities and individuals to become more self-reliant in the face of hazards being now common in the western world, following decades of Governments' pledges to protect against all threats, turning populations away from the 'nanny-state' (de Botton, 2011) is a challenge requiring significant time and effort on the part of Government's and key agencies (e.g. Weiss et al, 2011). Often places where hazards occur as disasters relatively infrequently are the least prepared for them. In New Orleans, the way in which authorities dealt with the aftermath of Hurricane Katrina (2005) has been closely examined and the survivors were 'controlled as criminals rather than assisted as victims' in the aftermath (Miller, 2011) in contrast with the way that the Brisbane flooding was dealt with in 2011 where the spirit of 'mateship' between individuals was very evident.

But these are not necessarily global issues. In places such as southern Europe and the tropics where intense rainfall is more prevalent than in northern Europe, greater account is taken of flooding in the design of buildings and their surroundings and in many places heavy rainfall occurs regularly in monsoon periods. This often makes buildings more resistant and people more resilient to flooding. But in many of these countries insurance is not available in the way in which it is in the western world, albeit there are major differences between approaches in each EU country as illustrated in Table 3 (modified from Botzen, 2010).

Spreading risks occurs at different levels (multi-layered) and as well as local (property and individual levels) catastrophe bonds are now used enabling a company, international organization or a government to issue bonds to protect themselves against predefined risks. Over 160 "catbonds" have been issued to date around the world to protect against pandemics, terrorism and natural disasters. (World Economic Forum, 2011).

Table 3.   Flood damage insurance in certain EU countries.

Arrangement	Netherlands	UK	France	Germany
Private insurance availability	No	Yes	Yes	Yes
Premium differentiation	No	Yes	No	Yes
Public reinsurance	No	No	Yes	No
Public compensation scheme	Yes	May be ad-hoc	The above is partly financed through taxes	Yes

4   EXAMPLES OF AN APPROACH TO MANAGING FLOOD (AND OTHER) RISK

Here two approaches are introduced that help to provide strategies to cope with hazards and risk. They address the problem of greater relative uncertainty that we now appear to face, mainly as a result of climate change as illustrated in Figure 3.

Hallegatte (2009) does not believe that increasing knowledge from now on will necessarily reduce the uncertainties illustrated in Figure 3. He advocates that users of climate change information need to change their approach to decision making to be more aligned with what they do for economic risks. This means that insuring against risks may be much more economic than providing hard, irreversible and expensive structural measures.

This aligns with the concept of a socio-technical system as being essential to support flood risk and response assessment (Gersonius, 2012). Socio-technical systems link physical (and non-structural) systems with actors (e.g., flood management organisations) and rules (e.g., acceptable standards) performing a particular function (e.g., flood risk management) (Geels, 2004). Different relations exist between the physical systems, actors and rules. The causal relations with respect to flood risk are well-understood, and these are generally described using the Drivers-Pressures-State-Impacts-Responses (DPSIR) concept. The system state includes the state of the physical flooding system, the actors and the rules. In any system state, the whole system has a quasi-stationary level of risk associated with it, where risk is considered as a function of the flood frequencies and impacts. Drivers, pressures and impacts are then considered in terms of how the system state may alter. Drivers and pressures act upon the system state, resulting in physical as well as socio-economic changes. This has both negative and positive effects on the level of risk associated with the system state, and this is related to the impacts. The impacts may lead to responses, which are diverse adaptations by the actors, many of which are non-structural.

Collectively, the above relates to the performance of the physical flooding system as well as the performance of the actors responding to flood risk which are indivisible. Next to causal relations, normative relations exist between the components of a system. A relation is normative if one component includes a rule which provides a structuring context for another component (Ottens et al., 2006; Birkmann et al, 2010). Rules can be categorised as cognitive, normative and formal (i.e., knowledge, behavioural norms and regulations, respectively). Actors use cognitive rules to shape perceptions of the future, and hence to make adaptation decisions in the present. Formal and normative rules also influence the behaviour and adaptation decisions of actors, as the actors are embedded in regulatory structures and social and organisational networks. Like the actors, the physical flooding system is structured by rules: for example, acceptable standards for flood risk management will limit the frequency or risk of flooding to a pre-defined level. The added value of using the socio-technical system concept is that the co-evolution of the technical system and socio-economic system, of structure and function becomes the focus of attention in risk and response assessment (Geels, 2004). From this viewpoint, the interactions within the flooding system should be considered as a dynamic process of mutual adaptations and feedbacks between the physical flooding system and the actors being impacted upon by flooding or responding to flood risk.

28

Figure 3. Diagrammatic illustration of humanity's understanding of natural hazards.

5 SUMMARY

Natural hazards seem to be increasing in frequency of occurrence and in impact. Events described as catastrophic also seem to be increasing year on year. Following the realisation that the earth's climate is changing, hitherto confident predictions about the variability in natural phenomena based on increasingly sophisticated measurements over the past century are now back into the 'uncertain' category. There is also a new awareness that natural hazards are often prompted from or exacerbated by human activity and that these are likely only to become more uncertain and cause bigger impacts. The most recent understanding of both the risks from natural hazards and what to do about these sees the integrated management of socio-technical systems as essential, rather than the old ways of dealing with natural threats by building ever bigger and better infrastructure; itself leading to even greater climate change.

ACKNOWLEDGEMENTS

The authors are engaged in several EU and UK funded programmes: North Sea Region INTER-REG IVb projects MARE and SKINT; North West Europe INTERREG IVb project Flood-ResilienCity; EU 7th environment framework programme PREPARED project; UK Flood Risk Management Research Consortium; UK Construction Industry Research and Information Association project on retrofitting surface water management systems. Work in all of these has contributed to this article as has discussions with the many colleagues involved in them.

REFERENCES

Ashley RM., Blanksby JR., Cashman A. Adaptable Urban Drainage - Addressing Change In Intensity, Occurrence And Uncertainty of Stormwater (AUDACIOUS) (2008). Main summary report. September. http://www.sheffield.ac.uk/penninewatergroup/publications/reports2.html

Barr S., Shaw G., Coles T. (2011) Times for (Un)sustainability? Challenges and opportunities for developing behaviour change policy. A case-study of consumers at home and away. Global Environmental Change.

Berkes, F., Folke, C., Eds. (1998) Linking Social and Ecological Systems: Management Practices and Social Mechanisms for Building Resilience. Cambridge University Press, Cambridge, UK.

Birkmann J., von Teichman K. (2010) Integrating disaster risk reduction and climate change adaptation: key challenges—scales, knowledge, and norms. Sustain Sci 5:171–184. DOI 10.1007/s11625-010-0108-y.

Botton de A. (2011) A Point of View: In defence of the nanny state. BBC mobile News magazine. 4th Feb. http://www.bbc.co.uk/news/magazine-12360045 [accessed 31-10-11]

Botzen W.J. (2010) Economics of Insurance against climate change. PhD thesis Vrije Universiteit Amsterdam. ISBN 978-90-8659-417-7

Dawson R., Ball T., Werritty J. et al (2011). Assessing the effectiveness of non-structural flood management measures in the Thames Estuary under conditions of socio-economic and environmental change. Global Environmental Change 21.2: 628–646.

Defra (2011) Flood and Coastal Resilience Partnership Funding policy statement on an outcome-focused, partnership approach to funding flood and coastal erosion risk management. 23 May. Department of Environment Food and Rural Affairs. Crown Copyright. http://www.defra.gov.uk/environment/flooding/funding-outcomes-insurance/

EA 2011. Environment Agency Flood Pages Home http://www.environment-agency.gov.uk/homeandleisure/floods/default.aspx [accessed 31-10-11]

EU, 2007, Directive 2007/60/Ec Of The European Parliament And Of The Council, of 23 October 2007 on the assessment and management of flood risks.

Evans EP., Ashley RM., Hall J., Penning-Rowsell E., Sayers P., Thorne C., Watkinson A. (2004). Foresight. Future Flooding Vol II – Managing future risks. Office of Science and Technology. April.

Everard, M. (2011). The Mayes Brook restoration in Mayesbrook Park, East London: an ecosystem services assessment. Environment Agency, Bristol.

Foley JA., Coe MT., Scheffer M., Wang G. (2003). Regime Shifts in the Sahara and Sahel: Interactions between Ecological and Climatic Systems in Northern Africa. Ecosystems 6: 524–539 DOI: 10.1007/s10021-002-0227-0.

Fuchs S., Kuhlicke C., Meyer V. (2011) Editorial for the special issue: vulnerability to natural hazards—the challenge of integration. Nat Hazards 58:609–619 DOI 10.1007/s11069-011-9825-5.

Ganzey LA., Razjigaeva NG., Grebennikova TA. (2011). Influence of natural catastrophes on the development of Southern Kuril Island landscapes in the Holocene. Quaternary International 237 15e23 http://dx.doi.org/10.1016/j.quaint.2011.01.007

Geels, F.W. (2004) From sectoral systems of innovation to socio-technical systems:: Insights about dynamics and change from sociology and institutional theory, Research Policy, 33, 897–920.

Gersonius B (2012). The Resilience Approach: To climate adaptation applied for flood risk. PhD thesis TUDelft. ISBN 978-0-415-62485-5.

Gersonius B., Ashley R., Pathirana A., Zevenbergen C. (*in press*). An identity approach for assessing the resilience of water and flooding systems to climate change. Natural Hazards.

Gunderson L. (2010) Ecological and human community resilience in response to natural disasters. Ecology and Society 15(2): 18.

Hallegatte S. (2009) Strategies to adapt to an uncertain climate change. Global Environmental Change 19 240–247.

Investopedia (2011) http://www.investopedia.com/terms/c/catastrophe-insurance.asp#axzz1cMOKGNqi [accessed 31-10-11]

ISO (2009a) ISO 31000, Risk management—Principles and guidelines, 2009, International Organization for Standardization, Case postale 56, CH-1211 GENEVA 20, Switzerland.

ISO (2009b) ISOGuide 73, Risk management—Vocabulary, First edition, 2009, International Organization for Standardization, Case postale 56, CH-1211 GENEVA 20, Switzerland.

Kousky C. (2011). Managing Natural Catastrophe Risk: State Insurance Programs in the United States. 5.1 (2011):153–171.

Krausmann E., Renni E., Campedel R., Cozzani V. (2011). Industrial accidents triggered by earthquakes, floods and lightning: lessons learned from a database analysis. Nat. Hazards. 59:285–300 DOI 10.1007/s11069-011-9754-3.

Kropp JP., Walker G., Menoni S. et al (2011). Risk Futures in Europe: Inside Risk: AStrategy For Sustainable Risk Mitigation. Springer. DOI: 10.1007/978-88-470-1842-6_5.

Kuhlicke C., Scolobig A., Tapsell S. et al (2011). Contextualising social vulnerability: Findings from case studies across Europe. Nat Haven't 58: 789–810 DOI 10.1007/s11069-011-9825-5.

Michel-Kerjan E. & Kunreuther H. (2011) Redesigning Flood Insurance. Policy Forum 22 July, Vol 333 Science. www.sciencemag.org

Miller LM. (2011). Controlling disasters: recognizing latent goals after Hurricane Katrina. Disasters. doi:10.1111/j.1467-7717.2011.01244.x.

Munich Re (2011). http://www.munichre.com/en/media_relations/press_releases/2011/2011_01_03_press_release.aspx [accessed 31-10-11]

NAO (2011). Flood Risk Management in England. National Audit Office. TSO, POBox 29, Norwich NR3 1GN. http://www.nao.org.uk/publications/1012/flood_management.aspx [accessed 31-10-11]

Newman R., Ashley RM, Molyneux-Hodgson S., Cashman A. (2011). Managing water as a socio-technical system: the shift from 'experts' to 'alliances'. Proc. Of the Institution of Civil Engineers. Engineering Sustainability. Vol. 164 Issue ES1. Paper 1000032. Doi 10.1680/ensu. 1000032. 95–102

Oberholzner F. (2011). From an Act of God to an Insurable Risk: The Change in the Perception of Hailstorms and Thunderstorms since the Early Modern Period. Environment and history 17.1 133–152.

Ottens, M., Franssen, M., Kroes, P., and Van De Poel, I. (2006) Modelling infrastructures as socio-technical systems, International journal of critical infrastructures, 2, 133–145.

Pantti MK., and Wahl-Jorgensen K. (2011). 'Not an act of God': anger and citizenship in press coverage of British man-made disasters. Media Culture & Society 33: 105 http://mcs.sagepub.com/content/33/1/105

Rahman SA., Walker W., Marchau V. (2008). Coping with Uncertainties About Climate Change in Infrastructure Planning – An Adaptive Policymaking Approach. Ecorys for: RAAD voor Verkeer en Waterstaat Rotterdam. ECORYSNederland BV, P.O. Box 4175 3006 ADRotterdam, Watermanweg 443067 GGRotterdam, The Netherlands.

Reid WV., Mooney HA., Cropper A. et al (2005). Ecosystems And Human Well-Being. Island Press. ISBN 1-59726-040-1.

Schad I., Schmitter P., Saint-Macary C. et al (2011). Why do people not learn from flood disasters? Evidence from Vietnam's northwestern mountains. Nat Hazards DOI 10.1007/s11069-011-9992-4.

Schmidt J., Matcham I., Reese R. et al (2011) Quantitative multi-risk analysis for natural hazards: a framework for multi-risk modelling. Nat Hazards 58:1169–1192 DOI 10.1007/s11069-011-9721-z.

United Nations (2011). World Economic and Social Survey 2011. ISBN 978-92-1-109163-2. eISBN 978-92-1-054758-1.

Vaught, S. (2009) "An "Act of God" Race, Religion, and Policy in the Wake of Hurricane Katrina." Souls 11.4:408–421.

Weiss K., Girandola F., Colbeau-Justin L. (2011). Les comportements de protection face au risque naturel: de la résistance à l'engagement [Protection behaviors with regard to natural hazards: From resistance to commitment]. Pratiques psychologiques 17 251–262.

World Economic Forum (2011). Global Risks 2011. Sixth Edition. An initiative of the Risk Response Network. January. ISBN: 92-95044-47-9 978-92-95044-47-0 REF: 050111 (http://reports.weforum.org/wp-content/blogs.dir/1/mp/uploads/pages/files/global-risks-2011.pdf)

Resilience and Urban Risk Management – Serre, Barroca & Laganier (eds)
© 2013 Taylor & Francis Group, London, ISBN 978-0-415-62147-2

The rise of resilience in large metropolitan areas: Progress and hold backs in the Parisian experience

M. Reghezza-Zitt
Ecole Normale Supérieure, Paris, France

R. Laganier
University Paris Diderot, Sorbonne Paris Cite, Paris, France

ABSTRACT: The analysis of flood risk and its management in Paris and its region (Ile-de-France) is a key of interpretation for understanding the evolution of management policies in the metropolitan areas. This chapter presents an historical approach of the flood risk management in Paris since the 19th century. It also gives an outlook on current management. Furthermore, we take a critical view of the practical implementation of resilience.

Paris and its region (Ile-de-France) face the risk of possible flooding, mainly due to the evolution of concerning metropolitan areas in general and their management policy. This results from the enormous growth in population and infra-structure in areas which have not been prepared for flood-risk (almost 1500 hectares of floodable areas have been used for construction between 1982 and 2008, and mainly in the close peripheral area.).

This risk has been increased by what can be referred to as a "metropolitan process" (Mitchell, 1995, Lacour and Puissant, 1999; Valache, 2003; Reghezza, 2006) This means areas witnessing a strong concentration of population, and the development of a centralized political, cultural and economical role of certain cities in a territory which is relatively small.

The potential damage is considerable: seventeen billion in material and an estimated twenty to forty billion in economic losses, according to data collected from the 1910 flood (IIBRBS *et al*, 1998). This will affect daily life, cause electrical cuts and set-backs in telecommunications, transport and fuel resourcing which will involve one to two million people, but also the major networks of the city (energy, water, transport). The functional risks do not necessarily depend on material damage, but also on the measures taken during the phase of alert. In such a context it will be necessary to close down the RATP network in order to block possible flooding and shut down the electrical implants for obvious security reasons. These diagnostic elements can be used to assess current standard methods used during risk management. With these elements, the aim is to mark certain reference-points by establishing the necessary steps which should be taken during an alert and the appropriate policy of prevention, in order to regain control of a situation in the least time possible and assure the functioning of certain services.

In other words, the study of this data shows how the concept of resilience can be considered as a new opportunity in risk management-logic, and the capability it has of modifying the general conception of risk (Pelling, 2003; Vale and Campanella, 2005; Veyret and Reghezza, 2006; Cutter, 2006; Dauphiné and Provitolo, 2007; De Bruijne et al, 2010).

1 THE PARISIAN EXPERIENCE: A PROGRESSIVE ADAPTATION OF SPECIFIC INSTRUMENTS USED DURING FLOOD ALERTS

Analyzing the evolution of practice during flooding in the Parisian region in a one hundred year time-span shows a progressive adaptation of technical instruments. Studying the flood of 1910 shows how efforts were mainly concentrated on minimizing damage (Ambroise-Rendu, 1997). Following the flood in 1876, works were done to lower water levels and reduce rising speeds, leading Parisians to believe that they were safe (works which have deepened the water bed of the branches linking l'Ile Saint-Louis and the Cité, the demolished the Monnaie barrier and the Ile Saint-Louis pier, modernized locks for navigation and re-profiled docks. We have also reconstructed certain bridges by using a minor number of piles and raised their level in order to help the flow of water). After the so-called "Great Flood", engineer Picard opted for a number of solutions as can be read in his report.

He speaks of the type of action which can be used to reduce the intensity of possible damage (mitigation, in its strict meaning) and actions to protect populations (the erection of barriers to contain potential over-spilling). The direct action taken to contain damage depends strongly on the conditions of drainage and the modification of the morphology of channels and banks so as to create drainage fields, and on reservoir stocks (this solution was proposed at the beginning of the XIX the century and was progressively used from the 1920s). From the 1960s onwards reservoirs blocks will turn out to be the key system. The system of protection consists of implanting defense mechanisms such as shutters, stop-gates, walls and dams which form a heterogeneous line of defense due to the fact that every area has different flood levels. For instance, certain districts hit by floods dating between 10 to 50 years, have a lower probability of risk in comparison to areas hit by floods dating over a hundred years.

In parallel to these procedures of protection and prevention of a potential crisis, decisions taken during the alert-phase play a fundamental role. In 1910 a series of measures had been taken for the following purposes: the evacuation of people in danger and protection of livelihood; the security of the population in flooded zones; assure the continuity of the city's infrastructure and services, assuring access to resources, waste and transport management, and handle the strategic functions of the city in order to prevent any potential political or economic risk. During the emergency the operational-team evacuated the population, while most of the waste was thrown to the Seine. A great of the frenzy was due to the population and the city, and its peripheral area had come to a standstill. Nevertheless, people were able to move by using horses or boats, and wood was used for heating; food was shared and the surveillance of neighborhoods was done by the population itself in order to discourage looting. The bourgeoisie was handled in a much more elaborate manner. The deputies being evacuated from the Bourbon Palace is charismatic, but we must also remember that the flood blocked the rail-road system and caused telephone cuts. The capital found itself to be isolated from two-thirds of the national territory, and from European countries such as the UK, Switzerland and Italy.

2   THE CONTEMPORARY STANCE, A COPERNICAN REVOLUTION?

The methods described in the 1910 flood were used up till the 1990s, due to several reasons. The absence of catastrophic events, apart from the floods in 1924 and 1955, had forged a risk-free general belief. Until this period there is an absence of conscience regarding potential dangerous situations. After the 1982 flood, the government understood that they needed to re-evaluate and reconsider what had been done until then. The techniques existing till then had given no useable data. It will take twelve years for researchers to write a paper assessing the events. After this many others will follow. All the people involved during the last flood in the series of interviews done in 2004 and 2006, all agree that there was a lack of preparation and experimented methods.

The 1990s mark a turning point, where we witness a progressive and important change in practice. This can be explained by various factors. From this period onwards, traditional methods which had focused on damage and protection are questioned by the academics, but also begin to be reconsidered on a national scale. The 1990s are indeed promoted by the UN as the "International Decade for Natural Disaster Reduction" (IDNDR). The IDNDR has given the risk governing-body new ideas, such as the concept of urban risk, the link between urbanism and risk, and the concept of vulnerability. Vulnerability questions the traditional viewpoint of risk, which has considered risk management as a series of technical solutions with the purpose of eradicating any possible damage and protect people and resources. This concept enables us to explain the political failure in regards to the topic and an inadequate financial backing and technical know-how. The techniques are efficient in reducing biophysical vulnerability, but are inefficient in terms of 'social vulnerability'. Vulnerability also gives great importance to population, considering them as active participants, and no longer a passive elements when facing a crisis. Vulnerability is thus an answer to an operational need which gives way to new solutions of management and questions the status quo (fig. 1).

France's situation is unique on an international scale. The several major catastrophic events which hit the country at the end of the 80 s brought to a change in crisis management. Government now concentrates on prevention, and the 1995 Barnier law is considered as a turning point in risk management. It has two main branches: prevention, as in the 1980 law, and a controlled plan of urbanization. Prevention leads to a risk culture, making people conscious of possible risks. Urban plans consider densely populated areas as highly potential risk zones due to the concentration of people and infrastructure, which higher the risk rate. Thus urbanization is seen as one of the factors causing damage to people and resources.

In the Parisian region area (Ile-de-France), as the rest of the national territory, we can observe a shift towards protection and prevention. The first studies of the 1990s are done to justify the financing of new barriers, but will soon be put aside because of the imparity between cost and earnings. The cost of estimated damage is so high that prevention seems necessary.

The solutions proposed are traditional ones: an appropriate training for the governing-body and leaders, and the stoppage of PPRIs (which today seems inefficient). The Ile-de-France context has an essential role: most areas are already constructed zones; many zones are strategic for metropolitan development and benefit from derogation. The strong taxation creates opposition and invalidates official documents. Prevention is reconsidered so as not to create unnecessary conflicts and not discourage potential investors. This system has shown to have also other solutions. The uniqueness of the French metropolitan area is its shift towards different paradigms of management when answering the need of such an exceptional region. The police working in the Defence area, near the administrative zone, in the 1990s was in charge of the flood assistance plan, which in time will progressively become the ORSEC. Nevertheless, the danger can no longer be faced with traditional techniques, inadequate when facing new potential situations of crisis. Actions of protection and prevention anticipating a potential risk are necessary, but this approach is not entirely sufficient and capable of resolving a situation of alert. Zero risk policy is impossible so it is more logical for us to move towards a policy of acceptance: minimize the submersions in order to reduce damage to material and consequent complications; concentrate state finance on security during emergencies, and promote autonomy by preparing the population for any eventual crisis and teach them how to protect their livelihood; make sure that certain urban strategic metropolitan

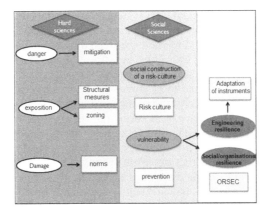

Figure 1.   Different concepts and approaches of risk management.

services will continue to function in order to be able to return to a "normal" situation. This is what is intended when professionals speak of resilience. The conversion towards resilience, mainly a term used by experts, is a pragmatic answer to an operational stalemate resulting from the change in the nature of a potential risk in the metropolitan Parisian area. The metropolitan risk is a hybrid one in which the governing—body is asked to act in a situation of uncertainty: a large-scale risk including material damage and important functional complications. The impact of an emergency has effects which go well beyond the area which is directly hit and rapidly spreads towards wider zones. Thus, the multiplicity of potential areas and people involved means that a traditional stance is no longer possible.

3 RESILIENCE, ON THE FIELD DIFFICULTIES

The practice of resilience today in the Parisian metropolitan area is translated in a series of pragmatic actions, such as plans of prevention and the elaboration of standard approach activities. Nonetheless, the strategy varies according to political will.

The companies in charge of infrastructure (like the RATP or the electric companies) have tendency to minimize the material and functional consequences of a major hydrologic event, which many today consider as inevitable. For this reason it is necessary to shift traditional methods towards techniques aimed at limiting damage and the duration of emergencies, and assure a rapid recovery of an ordinary situation. Possible action: preventive adaptation of the weak-points of a network by anticipating the effects of a crisis (Lhomme et al, 2010); a plan to manage an alert by blocking infiltrations and limit damage to material.

The industries belonging to the tertiary sector, which are less sensitive to the alteration of infrastructural material and more oriented towards strategies of adaptation, assure their functioning by transferring a certain number of their activities, delocalizing a number of traders, markets and banks. Even though innovative initiatives are taken to limit impacts and assure a certain continuity in economic activities, many basic conditions of resilience have not been prepared beforehand.

Resilience means considering the possibility of a catastrophic event and how we should come to terms with the utopian zero risk standing point (Laganier, 2006). This position means that technical solutions of protection are not enough. What's more, accepting the concept of resilience means that one must consider the complexity of urban systems and their multiple interdependencies and move away from a simplistic position. It also means developing a culture of communication between professions which are yet still too isolated behind linguistic and technical barriers. To these cultural deficits we can add a lack of efficient managing systems. There are still too many faults in regards to the preparation and training to face emergency and situations of alert, which could potentially lead to inefficient responses in extreme situations of danger (This means insufficient simulation and practice; no emergency plan; absence of a crisis cell and leaders which have not had the necessary preparation to face this type of situation; insufficient alert systems). These deficits are caused mainly by a predominant production-based society, which have superseded security meaning insufficient and inadequate communication and relations between different networks and organizations. The operations engaged following hydrologic events are often similar due to pragmatic reasons which are easier, faster, have a limited cost and are supported by the CatNat support system. They can also be considered to be somewhat symbolic because of the difficult task they have of (supposedly) completely erasing the misdoings of catastrophic events.

In the last two decades there have been strong efforts in urban renovation in new areas, but at the same rate strong territorial differences have been created in the region, meaning that different areas have different possibilities of reaction, depending mainly on their level of innovation and existing infrastructure. What's more, in spite of emergency plans of prevention in many enterprises, strong disparities remain in many other organizations present in the capital's region. A logic of management is being experimented but its practical and operational application still needs to be largely practiced and needs more on-the-field experimentation during a given situation of alert.

REFERENCES

Ambroise-Rendu M., 1997. *1910, Paris inondé*, Hervas, Paris, 111 p.

Cutter S. (ed), 2006. *Hazards, Vulnerability and Environmental Justice*, London, Earthscan Publishers.

Dauphiné A., Provitolo D., 2007. La résilience: un concept pour la gestion des risques, *Annales de Géographie*, n°654, p. 115–125.

De Bruijne M., Boin A., Van Eeten M., 2010. The rise of resilience, in *Designing Resilience. Preparing fo Extrem Events*, ed. Comfort L.K., Boin A., Demchak C., Pittsburgh, University of Pittsburgh Press, p. 13–32.

D'Ercole R., Thouret J.-C., 1996. Vulnérabilité aux risques naturels en milieu urbain: effets, facteurs et réponses sociales, *Cahiers des Sciences Humaines*, 96(2), p. 407–422.

Institution interdépartementale des barrages-réservoirs du bassin de la Seine (IIBRBS), Agence de l'eau Seine-Normandie, Ministère de l'environnement, Région Île-de-France, 1998. *Évaluation des dommages liés aux crues en région Île-de-France*.

Institution interdépartementale des barrages-réservoirs du bassin de la Seine (IIBRBS), Edater, 1998, Approche qualitative des impacts économiques des inondations sur le bassin de la Seine. Étude test sur le département du Val-de-Marne.

Gleyze J.-F., Reghezza, M., 2007. La vulnérabilité structurelle comme outil de compréhension des mécanismes d'endommagement, *Géocarrefour*, n°82, 1–2.

Lacour C., Puissant S., 1999. *La métropolisation. Croissance, diversité, fractures*, Anthropos, Paris, 196 p.

Laganier R., 2006. Territoires, inondation et figures du risque. La prévention au prisme de l'évaluation L'Harmattan, collection Itinéraire géographique, 257 p.

Lhomme S., Serre D., Diab Y., Laganier R., 2010. Les réseaux techniques face aux inondations ou comment définir des indicateurs de performance de ces réseaux pour évaluer la résilience urbaine, *BAGF-Géographies,* 2010-n°4, pp. 487–502.

Mitchell J.K., 1995. *Crucibles of Hazards: Megacities and Disasters in Transition*, United Nation University Press, New York, 535 p.

November V., 1994. Risques naturels et croissance urbaine: réflexion théorique sur la nature et le rôle du risque dans l'espace urbain, *Revue de géographie Alpine*, vol. 82, n°4, p. 113–123.

Pelling M., 2003. The Vulnerability of Cities: social resilience and natural disaster, Earthscan, London.

Reghezza M., 2006. Réflexions autour de la vulnérabilité métropolitaine: la métropole parisienne face au risque de crue centennale, Thèse de doctorat, Université Paris X - Nanterre, sous la direction d'Y. Veyret.

Valache M., 2003. *Entreprises et risques de crues à Paris et en Petite Couronne*, Chambre de commerce et d'Industrie de Paris.

Vale, J.V., Campanella, T.J. (eds.), 2005. *The Resilient City. How modern cities recover from disaster*, New York, Oxford University Press.

Veyret Y., Reghezza M., 2006. Vulnérabilité et risques. L'approche récente de la vulnérabilité, *Annales des Mines*, n°43, p. 9–13.

Resilience and Urban Risk Management – Serre, Barroca & Laganier (eds)
© 2013 Taylor & Francis Group, London, ISBN 978-0-415-62147-2

Flood-proof ecocities: Technology, design and governance

R.E. De Graaf
Deltasync BV, Delft, The Netherlands
Rotterdam University of Applied Sciences, Rotterdam, The Netherlands

ABSTRACT: Urbanization, land subsidence and other challenges create the urgent challenge for cities to transform into flood-proof ecocities. This chapter draws on the recent debate on future cities in the international water sector. Based on this debate, the chapter describes five characteristics of flood-proof ecocities. In addition, the needed governance elements for cities to transform into flood-proof ecocities are discussed.

1 INTRODUCTION

For the first time in history more people live in cities than in rural areas (UNFPA, 2007). Urbanization will continue to increase. It is expected that in 2050, 75% of the world's population will be living in cities (Arup, 2011). Population growth and urbanization mainly takes place in areas that are vulnerable to flooding such as coastal areas and river plains. In 2030, about 50% of the world's population is expected to be living within 100 kilometres of the coast (Adger et al., 2005). Land subsidence continues to threaten cities in delta areas. Examples of serious problems due to land subsidence are increased inundation frequency, increased flood impacts, salt water intrusion and groundwater nuisance. The main causes of land subsidence are groundwater overexploitation and groundwater drainage. In the Netherlands, pumping, drainage and land reclamation over the past centuries have resulted in a situation where a significant part of the country is located below the mean sea level. Many delta cities such as Jakarta, Bangkok, Shanghai and Venice face problems with regard to land subsidence. In Jakarta, land subsidence of 20 cm to 200 cm has been reported in various locations and land subsidence rates of more than 10 cm a year are not uncommon (Abidin et al. (2007). Extreme weather events increase both in nature and frequency. Data from Munich Re (2010) show a steady increase in the number of natural catastrophies from around 400 a year in the early 1980's to almost a 1000 a year in the late 2000's. Due to the rising number of extreme weather events and continuing urbanization in vulnerable areas, global flood damage in urban areas has also increased up to more than 100 billion US$ a year (Munich Re (2010). The IPCC (2011) predicts that climate change is likely to further increase the number of extreme weather events. Next to climate vulnerability, cities are vulnerable to globalizing networks of food and energy supply. Cities depend almost entirely on surrounding areas their water supply. Local water resources are hardly used. Due to economic development, land use changes and climate impacts, the water scarcity of urban areas will increase. Research shows that already today, water resources are overexploited in most river basins (Smakhtin et al. 2004). Another problem related to the water cycle is the depletion of nutrients, in particular phosphate. The global food supply depends on artificial fertilizers. The production of most fertilizers is based on mining of finite reserves of rock phosphate. Proven phosphate reserves are sufficient for 100 years of economic use (Driver et al., 1999; Isherwood, 2000). Therefore it is key that cities recover nutrients from wastewater streams to secure global food production.

2 CHARACTERISTICS OF FLOOD PROOF ECOCITIES

To deal with the problems that are outlined in the introduction section many scholars in the water sector have provided an outline of characteristics and properties of future cities. Important recent outlines with a strong connection to water management are Water Sensitive Cities (Wong and Brown, 2009), Cities of the Future (IWA, 2010), and Water Resilience for Cities (Arup, 2011). The main characteristics are summarized in Table 1.

From these visions on future cities and watermanagement follows that cities should be sustainable and they should be flood proof. Moreover they should have a positive influence on the environment and provide ecosystem services. To summarize these characteristics the concept of 'Flood Proof Ecocity' is introduced in this chapter as a concept for such a city. Such a city would have five basic properties that are discussed below.

2.1 Using water systems as a source

In flood-proof ecocities water systems are used a source for water, energy and nutrients. Urban water systems can function as a source for urban water supply. Instead of a waste, stormwater can be used as a valuable resource for water use functions that do not require the highest quality (Niemczynowicz, 1999). Fit-for-purpose water supply means that the functional use of water determines the quality of the water that is used and no longer drinking water is used for all purposes. This will increase the water efficiency of cities. It is estimated that China and India will need all runoff that is generated to meet urban and agricultural water demand in the next 20 years (Jury and Vaux, 2005). Urban demand from water supply catchments should therefore be reduced (Wong, 2006). Infrastructure that combines both centralised and decentralised water sources makes cities more flexible to adapt to external changes such as climate change. Gleick (2003) argued that community-scale, decentralised facilities must complement conventional centralised infrastructure.

Closing nutrient cycles of urban water systems through recycling is needed and the use of for transportation of waste should be abandoned (Berndtsson and Hyvönen, 2002). Some authors have therefore proposed urban agriculture as a function of urban water management

Table 1. Properties of future cities as described in literature.

Water resilience for cities	Cities of the future	Water sensitive cities
Increasing raw water storage capacity	Interconnected, localised communities	Cities as Water Supply Catchments: access to a diversity of water sources underpinned by a diversity of centralised and decentralised infrastructure
Combating salination	Compact liveable and sustainable cities	
Implementing water demand management	Resource neutral and harmonised with the environment	Cities Providing Ecosystem Services: provision of ecosystem services for the built and natural environment;
Improving river basin management	Sustainable cities as part of sustainable regions	
Deploying Water Sensitive Urban Design into city planning	Well managed water cycle	Cities Comprising Water Sensitive Communities: socio-political capital for sustainability and water sensitive decision making and behaviours.
Harvesting rainwater and recharging groundwater	All water is good water, fit-for-purpose	
Improving drainage networks	Water literate community involved in decision making	
	Information is accurate, useful and accessible	
	Adaptive, integrative policy, planning and leadership	
	Multi-faceted water management system	

(Larsen and Gujer, 1996; Niemczynowicz, 1999). This would allow for nutrient recycling on a local rather than on a global scale and would decrease the dependency of urban areas on the global system of food production.

Next to using urban water sytems as a source for water supply and nutrients, the can also function as a source of energy. Possibilities include the use of wastewater for biogas production (Davidsson, 2007), recovering heat from wastewater streams (Sukkar, 2011), and using the urban groundwater and surface water systems as a source of heat (De Graaf et al. 2008). In the Paleiskwartier in 's Hertogenbosch a pond is used to collect solar energy in summer. Heat from this pond is stored in an aquifer thermal energy storage and is used in winter for heating the houses, offices and apartments in the area (Aparicio, 2008).

2.2 *More functions for urban water*

Compact cities offer many advantages for sustainability including reduced transportation energy use, lower costs of infrastructure and improved competitiveness of public transport (e.g. Newman and Kenworthy, 1989). The International Water Agency has listed compact cities as one of the key properties of future cities (Table 1). At the same time, cities need to create more space for water and increase the water retention capacity to adapt to extreme weather events. Therefore, it is essential that multifunctional and flexible use of surface water space is promoted in flood-proof ecocities. An example of multifunctional use of the water surface is the construction of floating buildings (De Graaf, 2009). Floating urbanization enables multi-functional use of space in densely populated areas, without further increasing flood risk. During floods, floating constructions increase the coping capacity of an urban area. No damage to the construction will occur because floating buildings will adapt to the rising water level. In addition, these buildings may serve as emergency shelter during flooding. Because floating houses can be relocated, they are also flexible and reversible, which is a benefit to deal with uncertain future developments such as climate change.

Next to floating urbanization, another function of urban surface water that can be promoted is water based public transport. Cities such as Venice have already applied this concept for a long time. Historically, also Dutch cities used urban water systems intensively for transportation. In the Netherlands, water was the most important mode of transportation until the 19th century. In that period, the train became more important and many canals were filled in, because of hygienic problems and water pollution. However, in many cities the main water infrastructure is still present (De Graaf et al. 2010). In addition, many Dutch cities have plans to restore the historic water systems. This creates opportunities to use the water system again for water based urban transport.

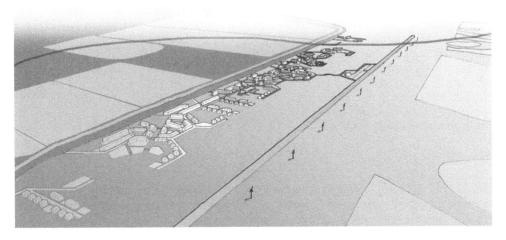

Figure 1.   Design of a floating district in Almere, The Netherlands.

2.3   *Involving citizens*

New technologies for water treatment, wastewater treatment, rainwater harvesting and energy production enable citizens to build, operate and maintain their own local water and energy supply. This can be done either at the individual scale or together with a group of neighbours. In the Netherlands, more and more spatial developments are based on collective private commissioning in which a group of citizens unite and fulfill the role of project developer. This creates opportunities for realizing self supporting blocks and houses. While this is all technically feasible, there are still many institutional barriers to implementation of such concepts. Water companies and energy companies increasingly use the term client, however there is usually no freedom of choice for citizens to buy services other than those provided by centralised water infrastructure. Citizen involvement in urban water management that goes further than paying taxes and fees is rare. Recent research in the Dutch water sector shows that urban water managers are convinced that involvement of citizens in urban water projects is key priority factor to achieve the objectives in urban water management (De Graaf et al. 2009). Private house owners will possibly have an incentive to cooperate in implementing local urban water solutions, if sewer and waterboard taxes are made dependent on the stormwater runoff and waste water production of private properties. In that case, water retention in a private garden leads to a lower water tax. This could stimulate local water retention and local water use for residential purposes. At the same time, it could create a market for local water management technologies. Also in reducing urban flooding vulnerability involving citizens can have benefits. Experience from Japan has demonstrated how better communication of flood risks to citizens can contribute to an improved coping capacity to deal with floods (De Graaf, 2009). Contrary to large scale centralized infrastructures, most local technologies should be adapted to local circumstances. Therefore, they require active participation of citizens as a source of context specific knowledge. Removing institutional obstacles for communities to be able to fulfill this role is essential to make this step. This will also create opportunities for innovative companies to develop small scale technologies for water management and energy supply.

2.4   *Water management initiative*

The role of water managers in spatial development processes should change from a reactive role at the end of the spatial development process toward a role where the water manager takes more initiative. By taking this approach, water resources and flood control interests can be included much earlier in the spatial development and urban design process. In the Netherlands, Waterboard De Stichtse Rijnlanden has taken such an initiative by developing a spatial plan for a flood-prood urban district. This district includes multiple flood-proofing concepts such as wet flood-proofing, dry flood-proofing, amphibious housing, floating housing, developing flood shelters and building on stilts.

Integration of water management in urban planning is an important prerequisite to implement sustainable urban water management solutions. In the city of Rotterdam, cooperation led to a change in stakeholder perception that facilitated the inclusion of innovations in urban planning policy (De Graaf and Van der Brugge, 2010). Transdisciplinary cooperation between spatial planners and water experts is also mentioned by other researchers as key condition for change in urban water management (e.g. Mouritz, 1996). Geldof (2005) proposes parallel plan making in which the knowledge of citizens and maintenance experts are incorporated in the design phase.

2.5   *Integration of flood control and urban planning*

In cities often space is lacking to adapt the urban environment to extreme weather events and a rising sea level. Increasing the height of dikes for instance, requires much space that would require demolishing many buildings. Therefore, integration of urban planning with flood control strategies is a key component of flood-proof ecocities. A good example of such a strategy is the development of superlevees in Japan (Stalenberg and Kikomori, 2008). Due to

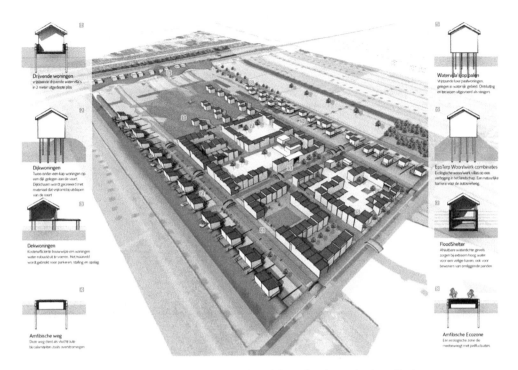

Figure 2.   Rijnenburg: Plan developed by waterboard for a flood-proof urban district.

the high value of buildings and infrastructure along the rivers in Tokyo and the importance of land ownership, it is not possible to demolish an urban area to construct a levee. Therefore, superlevees can only be created in combination with urban renewal projects. During the planning of an urban renewal project along the river, the river manager is involved and a superlevee is integrated in the urban renewal plan. Characteristic for this type of planning is the long term perspective. One segment of superlevee will have no impact on its own in reducing the flooding probability. This reduction will only be the case if multiple segments of superlevees are created and if these segments are connected in order to form a superlevee riverfront. This will take decades to accomplish. The example of the superlevee is therefore a good example of the value of integrating water management and urban planning, and the use of a long term perspective to transform a city in order to make it floodproof.

3   GOVERNANCE OF FLOOD-PROOF ECOCITIES

To be able to move from the current vulnerable and parasitic cities towards flood-proof ecocities requires system innovation rather than system optimization. This means a fundamental new way of working is needed that will be anchored in the mainstream practice of professionals and citizens in cities. Some essential elements for the transformation of current cities to flood-proof ecocities are outlined below.

3.1   *Stakeholder receptivity*

A crucial factor for implementation of innovation in mainstream practice is the receptivity of professionals and other stakeholders to new approaches, new policies and new technologies. The success of programs that aim to advance the application of a certain policy or technology depends not on the technical quality of these instruments but on the receptivity of the recipients of these change programs. For the full receptivity continuum to be addressed it is

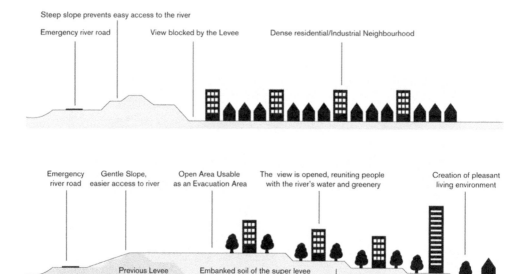

Figure 3.   Illustration of a conventional dike and superlevee.

necessary that urban water management practitioners (1) are aware of innovations, (2) want to apply these innovations, (3) have the required capabilities, and (4) have sufficient incentives to change their way of working (Jeffrey and Seaton, 2003). The mainstreaming process should be supported by the socio-political context. This includes the active development of values of organisations and regulations.

3.2   *Improving innovations*

Innovations in water management, energy and construction that require a new way of working, new skills and new knowledge will not easily be adopted even if they are technically and economically feasible. Innovations are often isolated showpieces that hardly contribute to the overall transformation of the urban water system (Brown, 2005: Hunt and Rogers, 2005). These innovations are hardly evaluated and improved. Neither does replication of demonstration projects take place on a large scale. Therefore, they remain isolated and fail to influence mainstream day-to day urban water management practice. Brown and Keath (2008) have argued that sustainability practitioners and strategists should focus more on providing the capacities and tools for replication and improvement of demonstration projects rather than just the demonstration of technology itself. For a successful transition from current cities to flood-proof ecocities, it is essential to make innovations more competitive compared to mainstream practices.

3.3   *Creating a commercial market*

It is necessary to create a commercial market for innovations in order to realize flood-proof ecocities. At this moment, there are hardly any incentives for developers and citizens to demand local water management and energy innovations. Such incentives could be created by awards, subsidies, increased competition among developers, and binding targets and regulations with regard to water management for instance on water robust buildings, source control, water quality, quality of the urban landscape, and integration with water management. For citizens, waterboard taxes should be made dependent on the surface of connected paved area to the sewer system to stimulate local water retention and water use on private property.

3.4 *The task of designers*

In a spatial development process with a more important role for citizens, the role of the designer will change from a determining role to a facilitating and inspiring role. Co-design with citizens becomes an important approach to develop solutions that are feasible from a societal point of view. The role of design is not to produce a blueprint for a future situation, but to inspire stakeholders to take the direction towards flood-proof ecocities. Additionally, the role of design is to provide input for discussions and stakeholder involvement.

3.5 *New roles for professionals and citizens*

The capacity of urban professionals and citizens to perform different roles than the traditional roles, is an enabling factor for realizing flood-proof ecocities. For example, using the urban water system for new functions implies that urban water managers and citizens will have to fulfill new tasks that are unfamiliar to them. An example of a new role is the waterboard as a developer of water plots for floating urbanization. The water utility company may become a facilitator of local water supply. A possibility is that local water treatments are owned by residents and that the water utility develops a new business model based on supplying technology and service contracts.

3.6 *Institutional mechanisms*

Professionals should no longer be rewarded based on effective execution of their fragmented statutory tasks, short term targets and costs minimisation. Instead, they should be judged on their contribution to the total system performance and long term targets. This creates room for stakeholders to be involved in long term collaborative projects. To secure public interests, selection of urban development partnerships should be based on costs, quality and system impacts rather than costs only. This should be supported by a management culture that is leadership driven rather than responsibility driven. Such a culture requires that professionals will have to do what is considered beneficial instead of doing what is legally prescribed. These changes will increase the potential of sustainability innovations to breakthrough to day-to-day professional practice.

4 CONCLUSIONS

Cities in delta areas are threatened by the impacts of climate change, urbanization and land subsidence. In this chapter, the concept of 'Flood-proof Ecocity' has been introduced to cope with the expected challenges and to summarize recent debates about future water cities. In a 'Flood-proof Ecocity' urban water systems are used as a source of energy, nutrients and local water supply. There is an important role for citizens as co-producers of the urban space. Citizens will also be involved in local water supply and energy production. Surface water in Flood-proof Ecocities is used for a wide variety of functions including floating buildings and water-based urban transport. The water manager is involved from the beginning of spatial developments. To adapt cities to extreme weather events and flood risk, flood control is integrated with urban development and urban renewal. Next to technical and design elements, this chapter has presented multiple building blocks that are needed for the governance of Flood-proof Ecocities. Important elements include: improving stakeholder receptivity, improving the competitiveness of innovations, creating a commercial markets for innovations, introducing new institutional mechanisms and facilitating new roles for citizens, technical professionals and designers.

REFERENCES

Abidin, H.Z., Andreas, H., Djaja, R., Darmawan, D., Gamal, M. (2008) *Land subsidence characteristics of Jakarta between 1997 and 2005, as estimated using GPS surveys.* GPS Solut (2008) 12:23–32.

Adger W.N., Hughes T.P., Folke C., Carpenter S.R. and Rockstrom J. (2005). *Social-Ecological Resilience to Coastal Disasters,* Science, 309, 1036–1039.

Aparicio, E. (2008) *Urban Surface Water as Energy Source & Collector.* MSc Thesis, Deltares/TU Delft, Delft, The Netherlands.

Arup (2011) *Water Resilience for cities. Helping cities build water resilience today, to mitigate the risks of climate change tomorrow.* Arup Urban Life Report.

Berndtsson, C.J. and I. Hyvönen (2002) *Are there sustainable alternatives to water-based sanitation system? Practical illustrations and policy issues.* Water Policy 4 (2002) 515–530.

Brown, R.R. (2005) *Impediments to Integrated Urban Stormwater Management: The Need for Institutional Reform.* Environmental Management, 36(3), 455–468.

Brown R.R and Keath, N. (2008) *Drawing on Social Theory for Transitioning to Sustainable Urban Water Management: Turning the Institutional Super-tanker*, Australian Journal of Water Resources. 12(2), 1–12.

Davidsson, A.(2007) *Increase of Biogas Production at Wastewater Treatment Plants Addition of urban organic waste and pre-treatment of sludge.*

De Graaf, R.E., F.H.M. van de Ven, I. Miltenburg, G. van Ee, L.C.E. van de Winckel en G. van Wijk (2008) *Exploring the Technical and Economic Feasibility of using the Urban Water System as a Sustainable Energy Source.* Thermal Science Vol 12, No 4, pp 35–50.

De Graaf, R.E., R.J. Dahm, J. Icke, R. Goetgeluk, S. Jansen and F.H.M. van de Ven (2009) *Receptivity to transformative change in the Dutch urban water management sector.* Water Science and Technology Vol 60, No 2, pp 311–320.

De Graaf, R.E. (2009) *Innovations in urban water management to reduce the vulnerability of cities. Feasibility, case studies and governance.* PhD thesis, Delft University of Technology.

De Graaf, R.E. de and Van der Brugge, R. (2010). *Transforming water infrastructure by linking water management and urban renewal in Rotterdam*, Technol. Forecast. Soc. Change (2010), Vol77, 8, pp 1282–1291.

De Graaf, R.E., L. Dietz, K.M. Czapiewska, W. Lindemans, and B. Roeffen (2010) *Water voor Bereikbaarheid. Het gebruik van historische waternetwerken voor het verbeteren van de bereikbaarheid van Nederlandse binnensteden.* Stimuleringsfonds voor Architectuur, Rotterdam, Nederland.

Driver J., Lijmbach D. and Steen I. (1999) *Why Recover Phosphorus for Recycling, and How?* Environmental Technology 20(7), 651–662.

Geldof, G.D. (2005) *Coping with complexity in integrated water management. On the road to Interactive Implementation.* Tauw. Deventer. The Netherlands.

Gleick, P.H. (2003) *Global freshwater resources: soft-path solutions for the 21st century.* Science 302 (5650), 1524–1528.

Hunt, D.V.L. and Rogers C.D.F. (2005) *Barriers to sustainable infrastructure in urban regeneration, engineering sustainability.* Proceedings of the Institution of Civil Engineers, 158, (ES2), 67–81.

IPCC (2011) *Special Report on Managing the Risks of Extreme Events and Disasters to Advance Climate Change Adaptation.*

Isherwood, K.F., (2000) *Mineral fertilizer use and the environment.* International Fertilizer Industry Association/United Nations Environment Programme, Paris, France.

International Water Association (2010) *IWA Cities of the Future Program Spatial Planning and Institutional Reform Discussion Paper for the World Water Congress*, September 2010.

Jeffrey, P. and Seaton, R.A.F. (2003) *A conceptual model of 'receptivity' applied to the design and deployment of water policy mechanisms.* Environmental Sciences 1, 277–300.

Jury W.A. and Vaux Jr H. (2005) *The role of science in solving the world's emerging water problems.* Proceedings of the National Academy of Sciences 102, 15715–15720.

Larsen, T.A., W. Gujer (1996) *Separate management of anthropogenic nutrient solutions (human urine).* Water Science and Technology 34(3–4): 87–94 s.

Mouritz, M. (1996) *Sustainable urban water systems: policy and professional praxis.* PhD thesis Murdoch University, Australia.

Munich Re (2010) *Münchener Rückversicherungsgesellschaft Georisk Research.*

Newman, P.W.G. and Kenworthy, J.R. (1989). *Gasoline consumption and cities. A comparison of U.S. cities with a global survey.* APA Journal 1989. 24–37.

Niemczynowicz, J. (1999) *Urban hydrology and water management- present and future challenges.* Urban water, 1, 1–14.

Smakhtin, V, Revenga C. And Döll, P. (2004) *Taking into Account Environmental Water Requirements in Global-scale Water Resources Assessments.*

Stalenberg, B. and Kikumori, Y. (2008) *Urban flood control on the rivers of Tokyo metropolitan.* pp. 119–141. In: Graaf, R.E. de, and F.L. Hooimeijer (2008) Urban Water in Japan. Urban Water Series volume 11. Taylor and Francis, London, UK.

Sukkar, R(2011) *Harvesting energy out of wastewater.* Presentation International Water Week Amsterdam.

UNFPA (2007), *State of World Population 2007*, ISBN 978-0-89714-807-8, United Nations Population Fund, New York, USA.

Wong, T.H.F. (2006) *Water Sensitive Urban Design*, The Journey thus far. Australian Journal of Water Resources 10, 213–222.

Wong T.H.F. and Brown R.R. (2009) *The Water Sensitive City: Principles for Practice*. Water Science and Technology, 60(3), 673–682.

The resilience engineering offer for municipalities

J.P. Arnaud
Egis, Lyon, France

M. Toubin
Egis, France
University Paris Est—EIVP, Paris, France
University Paris Diderot, Sorbonne Paris Cité, Paris, France

C. Côme
Egis, Lyon, France

D. Serre
University Paris Est—EIVP, Paris, France

ABSTRACT: This article aims at demonstrating the interest of the resilience approach in risk management and urban planning. We first describe the French regulatory framework in risk management, the latest evolutions and its drawbacks in making cities both resilient and sustainable. Indeed, we argue that local authorities lack technical skills, organizational powers and resilience culture in promoting an integrated management though the latest French and European strategies are pushing in this direction. Given this context, the Resilis project, funded by the French National Research Agency and lead by Egis, aims at providing a new offer for municipalities willing to assess and improve their territory's resilience. The last part of this article thus describes the method and tools developed to help local authorities in meeting the need for an integrated management of resilient cities.

1 INTRODUCTION

Every year, many communities are disrupted by disasters and have to face human losses, important damages, reconstruction challenges and population expectancies. Even more communities are exposed to risk and have to cope with development constraints, loss of attractiveness and population demand in protection. Then local authorities are confronted to several issues sometimes contradictory, but they lack a global understanding of issues interaction, of short and long term consequences (Boin et al. 2010). The regulatory framework confirms this sector-based consideration of development and risk issues, so that local authorities need a new approach to overcome the limits of the existing tools. Resilience can be seen as a new paradigm in risk management insofar as it enables an integrated analysis of systems like cities, acknowledging possible disturbance and promoting sustainability (Ahern 2011). Then a systemic analysis of cities takes into account interactions and retroactions, spatial and temporal scales in order to make it more resilient (Godschalk 2003).

The first part will provide a short overview of the regulatory framework of France in order to understand the shortcomings of risk management nowadays. This fragmented legislation is not specific to France, thus the new issues identified in the second part are not limited to France. Finally, the new resilient approach is described in the third part and operational tools to improve it are sketched by the Resilis Project outcomes.

2 RISK MANAGEMENT AND URBAN DEVELOPMENT FOR FRENCH LOCAL AUTHORITIES

2.1 *A traditional top-down approach based on engineering*

Because of historical reasons, which have already been analyzed in other works (Veyret & Reghezza 2006), France's risk management system heavily relies on the government and its engineers. In the past, solutions to protect populations from risks (mainly natural or technological) were basically structural and shaped the territory landscape with infrastructures, river management structures, non-built areas. The expertise knowledge of Engineers Corps was often forced on the local communities in order to avoid damages, notwithstanding adaptation strategies that populations had already developed to live with the risk for long. This conflict between risk management and local development is still paramount nowadays though the national government has decentralized many of its former prerogatives and gave more powers to local authorities (see 2.2). Recently, the ministry in charge of risks prevention has also discussed integrated management with local authorities, public establishments and private companies during a symposium on Natural Risk Management.

Albeit questionable and questioned, the central role of the French government ensures that public interest and territorial equity are respected when managing risks. The French government gives the main directions for risk management, it combines 2 strategies: prevention and intervention. The approach is comprised of 4 steps: 2 in each strategy. The first one is knowledge and risk analysis that requires being able to observe, model or assess potential risks. Government agencies are responsible for this first task. The second one is land use management and is to be implemented locally by municipal rules following the national directives. Crisis management is the third step, which concerns intervention. Here is a major shortcoming in the French risk management process because several actors are responsible for crisis management, at different scales depending on the event extent. Last of all is preventive information that is under the responsibility of the state when it concerns information of local authorities (mainly about risks) and under the responsibility of local authorities when it is about population information (mainly about safety instructions).

Figure 1 shows the risk management hierarchy from the state to the municipality, each official responsible for each scale and the main document regulating either prevention (on the left hand side) or intervention (on the right hand side).

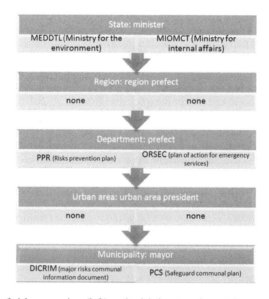

Figure 1.   Hierarchy of risk prevention (left) and crisis intervention (right).

Almost no connection is made between risk prevention and intervention, the consequence is that reconstruction is scarcely considered prior to an event. It results in decision-making, during the crisis or in the aftermath, that is not adapted to reduce vulnerability to future events, or when this aspect is taken into account but not prepared, in hindered decision-making. Risk management should be integrated in the urban development framework from readiness to reconstruction and maybe within the hands of one coordinator only.

Indeed, the second drawback identified in Figure 1 is the multitude of territorial scales involved in risk management. Some competences are in charge of the department level and other of the communal level, but other scales are neglected albeit means and resources are shared among all scales as well. The most important dichotomy is highlighted by the urban area scale which is more and more often in charge of urban planning but has no power in risk management. As above-mentioned, these hindrances are more and more taken into account within the new national strategies concerning risk management.

2.2   The risk management, a sector-based responsibility

If urban planning is to be consistent with risk management measures, those two responsibilities don't lie in the hand of the same institution. The need for multiple measures to counter risks at different scales has led the French legislation to multiply the documents enforcing risk management. Indeed, the French legal system relies on several Codes in which laws have been added over time to meet the need for risk management. As a caricature, one could say that, after each disaster, laws have been enforced to prevent another similar disaster (Ramroth 2007). Risk management is said to be comprised of a sum of policies dedicated to each actor (Figure **2**). Five major codes rule risk management, planning, construction and security issues.

Besides national rules and laws, the main tool to plan for risk management is the Plan for Risk Prevention (PPR) implemented by governmental services and often forced on local authorities without their consultation. Examples of conflicts around PPR implementation, trials or years of negotiation, but also permissive or inconsistent documents are numerous (Martin et al. 2010). Moreover, the lack of risk management is often pointed out when a disaster occur and everyone realizes nothing had been done to mitigate the consequences... And yet, this document is supposed to be translated into local planning document, so that future developments respect the analysis concerning risk exposure, described in the PPR. In doing so, urban development should be consistent with risk mitigation, but at the same time urban development can be hampered by risk mitigation (Faytre 2005). Thus representatives are often reluctant to implement those constraining measures on their territory.

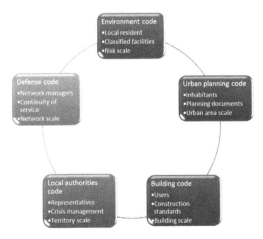

Figure 2.   Risk management, a sum of policies at different scales.

The issue is even more complicated by the fact that the planning competence is often transferred at the urban area scale, though risk prevention is always defined at the communal scale. To add another territorial scale, risk identification and management (concerning rivers mainly) is usually more relevant at the catchment scale, mountainous massif or geographical unity. In France, public institutions are in charge of water management at a completely different scale than departmental or regional scale. It enables managing flood risk outside urban areas but it requires collaboration, involvement and resources of many local authorities or private actors. Those institutions are the one that can foster an integrated risk management approach. That is why they are currently developing new tools, according to the European Flood Directive, that seem more adapted to the new issues in risk management and urban development (cf. 3.2).

With this short presentation of the French regulatory context, it is obvious that it is way too much fragmented to ensure a consistent risk management. Many difficulties stream from this sector-based approach and many mistakes are enabled by the vagueness of certain responsibilities. When sustainable development is supposed to foster a global approach of our world development, it seems that risk management, as is now, cannot entail sustainability.

3 LIMITS AND NEW REGULATIONS TO OVERCOME IT

3.1 *The subsequent limits of this strategy*

The top-down approach and the multiplicity of actors, responsibilities and scales created by the juxtaposition of rules poses several limits to risk management. First of all, it induces a certain confusion in risk responsibilities because many tools and regulations exist at different scales, under the responsibility of different actors, sometimes at different phases: decision, implementation, management,... For a private owner, a new resident or a private firm, it's not easy to find good information, advice and then help (technical or financial) to take risks into account in his development. Conflict between risk issues (security) and local development is paramount here. Local representatives have to play with real estate pressure, territory attractiveness, increasing costs, diminishing means and resources in order to meet the need for both security and development. That is why they need to be accompanied in the risk mitigation process. They need to be able to understand the effects and consequences of protection measures, laissez-faire policies or resilient solutions (cf. part 4).

Different parallel issues are also playing a part in this complicated problem. The most obvious one is sustainable development because when considering developing cities, economical, social and environmental sustainability are to be assessed. Indeed risk management is not just improving security; it is also fostering urban development with safe activities, quality of life and respected environment. Those competences are under the responsibility of several actors but the municipality should guarantee the harmonious articulation of all issues. Local authorities have to take into account new evolutions in economy, technology, social expectancies or environmental changes in order to provide an adapted answer. Then, sustainable development recommendations and risk mitigation policies should be assessed together with an integrated view of the urban system.

In this conflict analysis between development and security, people have a lot to say and the recent increase in public participation should support this involvement of people in what will shape their future life or city. That's why new requirements, particularly from the European Union, are giving more importance to public participation in environment management.

3.2 *The latest development concerning risks and networks*

The European Flood Directive sets up new requirements in national risk management policy. The main evolution is the definition of the risk reference: the moderate risk shall now be the 100-year flood, though the French legislation, for instance, considered this level as a major risk. In practice, very few French local authorities have planned to protect beyond the

100-year flood. Then the whole regulatory dispositive is to be adapted to these new expectancies. One of the tools to adapt the Flood Directive in France is the program for flood risk preventive actions plan (PAPI). It aims at embedding flood risk management in the local context, at the catchment scale, optimizing the resources and articulating the approach with the former risk management documents. This kind of tool is likely to be much better adapted to meet the needs for more consistency and simplicity. Then every stakeholder involved in risk management has to comply with new requirements that will foster an integrated risk management, and not limited to flood risk.

Networks managers have been lately identified as key stakeholders in risk management. Indeed they are responsible for several urban services that are essential to the city: power supply, drinkable water, sewage, telecommunications, and transportation. Since the Civil Security Modernization Law in 2004, continuity of service is required for those functions as far as they play an important part both during the crisis and in the aftermath for recovery tasks. The national Defense Code also enforces new requirements upon "vital activity sectors" so that security and continuity issues are now taken into account along with competitiveness and/or public interest. It's important to emphasize here that even if many public services are delegated, the local authority remains responsible for it, and in particular in case of crisis. The main hindrance in complying with these issues is not the ability of a manager to operate its system properly. In fact, networks are dependent from each other and a failure on one network could impact the others as well. And yet, statutory requirements are not taking interdependencies into account. One network could be highly reliable intrinsically, if it doesn't have electric power to operate its equipment, the service won't be delivered.

All the issues aforementioned emphasize the lack of integration in risk management: several legislations are addressing different issues; all territory scales are involved but maybe not at their full range of capacity; new (European) regulation seems to go in the right direction but still add another layout and last of all, specific requirements on technical networks are relevant but not completely efficient. Thus, it is necessary to develop a new approach able to tackle this lack of integration; as seen by the government, the resilience concept might be a promising tool.

4 THE RESILIENT APPROACH AND ITS OPERATIONAL TOOLS, A NEW OFFER TO IMPROVE SUSTAINABILITY

4.1 *How useful is the concept of resilience?*

First derived from physics, resilience is now broadly used in many disciplines among which ecology. This concept enables the assessment of the capacity of a system to face and then recover from shocks, whether using its intrinsic absorption capacity, reorganizing (adapting) its components or benefitting from its relations with other systems in order to find a new equilibrium state (Holling 1973). Thus, a systemic approach of the behavior of a system coping with a disturbance is facilitated by this global approach that takes into account short and long-term evolutions.

A city can be seen as a system comprised of several components interacting with each other (Figure 3) in order to perform the different functions of the city: production, employment, decision-making, care, education, information, entertainment...

Risks are likely to disrupt these functions which could have dramatic consequences on the economy, quality of life, attractiveness or sustainability of the city (Bruneau et al. 2003). That's why a systemic evaluation of the behavior of cities facing disturbances is needed. Resilience enables this systemic consideration of cities where each component contributes to the global resilience and where interactions are thoroughly taken into account. The aim is to assess the ability of the components to maintain urban functions whether resisting the disturbance, using redundant components or adapting functioning. Then urban resilience is defined as "the ability of a city to absorb a disturbance and recover its functions after the disturbance" (Lhomme et al. 2011).

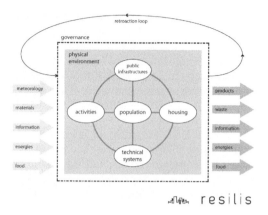

resilis

Figure 3.   The city as a system of 5 interacting components within a governance framework (Lhomme et al. 2011).

In risk management, resilience is an integrated approach to manage disturbances and their consequences on the urban system. It is not just focused on protection and crisis management but runs from prevention to recovery so that it can answer the continuity issues identified in 3.2. A multi-scale approach (both temporal and spatial) is also facilitated by the approach with the assessment of long-term impacts on the urban system, but also on the other systems in relation with it (other cities, environment providing resources) (Dauphiné & Provitolo 2007). Hence, this integrated approach seems able to gather all issues and tackle them within the same framework by a single authority that should probably be the city or the urban area. In doing so, we are likely to improve consistency within policies and foster more transparency in responsibilities and issues, which is a real shortcoming of current approaches (3.1).

4.2   *Resilis, a research project to improve resilience*

Resilis is a research project aiming at developing operational tools dedicated to local authorities that would like to improve their community's resilience. In the first tasks of the project many new issues and needs emerged from the assessment of the city as a system of systems (workpackage 1) and the analysis of the legal framework (shortly summarized in part 2). Then it appears that cities and urban areas needed new approaches to improve their resilience to multi-risks situations, often managed by different actors with contradictory objectives. The project is based on several feedbacks (WP2) that enabled the identification of urban vulnerabilities (WP3). Then resilient characteristics are highlighted (WP4) and methods and tools are developed (WP5) and validated with two experimentation fields in France (WP6).

The tools have been developed around three axes in order to take into account each shortcoming identified in the former parts: technical solutions, organizational strategies and resilience culture promotion. These axes are gathered in a consistent approach dedicated to local authorities that would like to assess and improve their territory's resilience (Figure 4).

The general method is comprised of several tools. The first one is a consensus exploration tool that enables the identification and discussion of acceptable solutions facing one or several issues (Da Cunha et al. 2010). On the basis of solutions assessment (both economical and social), elected representatives and other stakeholders are able to express their preferences and confront their opinions so that they find a common solution. The second tool is dedicated to interdependencies between urban functions. On the basis of an auto-diagnosis performed by a network manager, dependencies to other systems are identified and prioritized. Then all diagnoses are put together to represent functional dependencies within the urban system in order to discuss interrelations, acknowledge criticality and discuss common solutions (Toubin et al. 2011). With this kind of tool, the usual sector-based consideration of networks resilience is overcame and the resilience of the global system of systems should be improved. But the auto-diagnosis tool is not sufficient to realize a detailed assessment of the

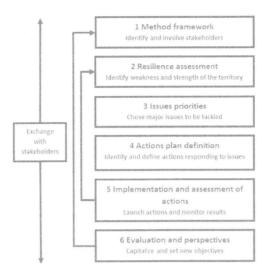

Figure 4. The Resilis method to foster an integrated management of urban resilience.

city functioning. More information concerning network components, failures localization and then issues impacted is needed. That is the point of the third tool which should be a GIS tool to model all networks, locate populations, critical infrastructures, economic activities and other decision-making centers in order to help the local authority prioritizing means and resources in improving resilience (Lhomme et al. 2011).

All along the project, recommendations concerning good practices in improving resilience are also identified. They will be provided to local authorities interested in this approach so that recommendations are adapted to the local context and issues. Indeed, regulatory evolutions, new requirements in contracts, needs for information, education or communication toward populations should be developed according to the priorities identified by the experimentation of the tools above-mentioned.

5 CONCLUSION

Urban resilience improvement is now confirmed as a powerful tool to manage risks within a community and within the regulatory framework that needs consistency and simplicity. Our resilient approach is based on technical networks resilience, which is already identified as a paramount issue in economic defense. Indeed, several regulations enforce continuity requirements upon urban services in order to maintain the city's functioning and improve adaptive capacity when facing disturbances. A complete analysis of networks behavior and interdependencies should be enabled within a GIS tool but several limits hamper its development. First of all, managers are reluctant to share information concerning their infrastructures and equipments, because of security reasons. New French regulations are being enforced in order to make up for this lack of data by forcing managers to share their data. By the time it is really applied, other tools can be used to identify interdependencies and raise awareness of network managers concerning resilient issues. The Resilis project will focus on these tools in order to validate the approach with local authorities and make sure the tools are answering their needs and difficulties.

Then with a consistent methodology comprised of several specific tools to assess systems resilience, the Resilis project accompanies local authorities in improving their resilience. This method could be seen as a way to give more consistency to urban policies, in fostering collaboration and exchanges between decision-makers, managers and populations. In doing so, local authorities are likely to improve their sustainability as well, so that the concept finds in the resilience approach a way to be implemented within cities.

ACKNOWLEDGEMENTS

Resilis (www.resilis.fr/en) is a project funded by the French Research National Agency (ANR-09-VILL-0010 VILL).

The authors address their thanks to the Resilis partners.

Resilis gathers researchers from the private sector (Egis, Iosis, Sogreah) involved in urban development or risk prevention, research laboratories (EIVP—Paris School of Engineering, LEESU—University of Marne-la-Vallée department of research in urban engineering, REEDS—University of Versailles-Saint Quentin department of research in economy, ecology and sustainable development), public research (Cemagref—risk assessment), and foundations (Fondaterra—foundation for sustainable territories).

REFERENCES

Ahern, J. 2011. From fail-safe to safe-to-fail: Sustainability and resilience in the new urban world. *Landscape and Urban Planning*, 100(4): 341–343.

Boin, A., Comfort, L.K. & Demchak, C.C. 2010. The rise of resilience. In Louise K Comfort, Arjen Boin, & Chris C Demchak, (eds.) *Designing resilience: preparing for extreme events*. Pittsburgh, USA, University of Pittsburgh Press: 1–12.

Bruneau, M. et al. 2003. A framework to quantitatively assess and enhance the seismic resilience of communities. *Earthquake Spectra*, 19(4): 733–752.

Da Cunha, C. et al. 2010. Analyse et discussion des résultats de l'évaluation finale—les avenirs de l'exploitation agricole de la Bergerie nationale, Rambouillet, France, REEDS, Université de Versailles Saint-Quentin-en-Yvelines (UVSQ).

Dauphiné, A. & Provitolo, D. 2007. La résilience: un concept pour la gestion des risques. *Annales de géographie*, 2007/2(654): 115–125.

Faytre, L. 2005. La prise en compte des risques majeurs en Ile de France: une composante indispensable de l'aménagement du territoire. *Cahier de l'IAU*, Les risques (142): 7–18.

Godschalk, D.R. 2003. Urban Hazard Mitigation: Creating Resilient Cities. *Natural Hazards Review*, 4(3): 136.

Holling, C.S. 1973. Resilience and stability of ecological systems. *Annual Review of ecology and systematics*, 4: 23.

Lhomme, S. et al. 2011. A methodology to produce interdependent networks disturbance scenarios. In ASCE (ed.) *International Conference on Vulnerability and Risk Analysis and Management*. University of Maryland, Hyattsville, MD, USA: 10.

Martin, B. et al. 2010. Territorialisation ou déterritorialisation du risque? Analyse comparative et critique de la procédure de réalisation des PPRNP. *Riséo*, 1: 83–98.

Ramroth, G. 2007. Planning for disaster: how natural and manmade disasters shape the built environment, Kaplan Publishing.

Toubin, M. et al. 2011. Improve urban resilience by a shared diagnosis integrating technical evaluation and governance. In *EGU General Assembly 2011. Vienna, Austria*.

Veyret, Y. & Reghezza, M. 2006. Vulnérabilité et risques—L'approche récente de la vulnérabilité. *Annales des Mines—Responsabilité & Environnement*, 43: 9–13.

Resilience and Urban Risk Management – Serre, Barroca & Laganier (eds)
© *2013 Taylor & Francis Group, London, ISBN 978-0-415-62147-2*

Strategies of urban flood integration—Zollhafen Mainz

C. Redeker
CO/R Cities On Rivers, Munich, Germany
Consultant Stadtwerke Mainz AG, Germany

ABSTRACT: Converting the inner city harbour Zollhafen in Mainz (Upper Rhine) into a new living district promotes a compact european city model by contributing to a decrease in land consumption and thus run-off. As a development outside of the municipal flood defense, it demands for flood adapted construction and a "retention-neutral" development to comply with German water legislation. The Zollhafen is being developed as a model project for flood adapted building in a partnership between the Stadtwerke Mainz Corporation (the municipal utilities company) as the owner of the harbour and the Ministry for the Environment of the State of Rheinland-Pfalz. Apart from defining requirements for building in the flood plain, different strategies and instruments are being developed to trigger innovative adaptive designs and to create public spaces which raise awareness for the inherent, but, due to the height of the site, rare flood risk.

1   BETWEEN MITIGATION AND ADAPTATION—THE COMPACT EUROPEAN CITY MODEL

Spatial Spatial flood management today involves two strategies, mitigating floods by giving more space to the river and adapting to floods by (re-)developing areas at risk to become flood-resilient. As developments outside of the municipal flood defense, harbour conversions demand for flood-adapted construction and a 'retention-neutral' development to comply with German water legislation. Converting the inner city harbour Zollhafen in Mainz on the Upper Rhine, traditionally positioned outside of the flood defense into a new living district is a typical flood adaptation measure.

In the 1990s, often competing with cities more privileged regarding topography and climate also in terms of branding, city centers and their representative water fronts were rediscovered. These sites were no longer developed by omnipotent governments, but market-driven actors and public-private partnerships (Meyer, 1999). Harbours have been converted into privileged urban developments for dwellings and office space paired with the creation of a new public realm. The 19th century promenade, before limited to the historical city, is now expandable along the full length of the compact cities urban river front. Due to an increasing flood risk and according changes in planning legislation, today the redevelopment of former harbours into new urban quarters also involves combined spatial and operative measures to manage that risk.

Beyond the project scale, inner city harbour conversions are part of a compact european city model as they promote density, short distances, and a public realm. As an alternative to further expansions into the periphery, redeveloping urban waste lands contributes to a decrease in land consumption. In this way, creating new dwellings on inner city sites coheres with the Aalborg Commitments 2004 and the Leipzig Charter on Sustainable European Cities 2007 and the ambitious aims of the german government to reduce land consumption from today 113 hectares to 30 hectares per day by 2020 (Bundesregierung, 2002). By redeveloping inner city sites, the creation of additional impervious surfaces and thus a further increase in run-off is avoided (see Bronstert et al, 2001 and Haase, Nuissl, 2006). Taking the regional

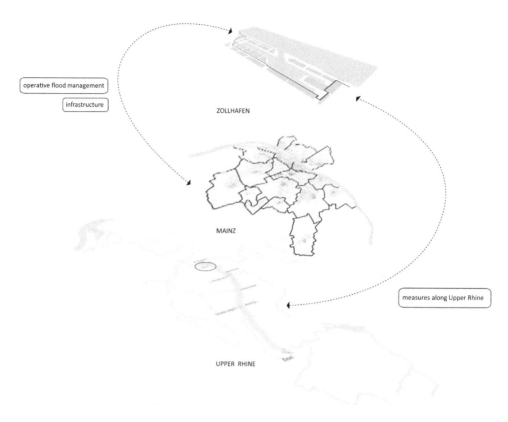

operative flood management

infrastructure

ZOLLHAFEN

MAINZ

measures along Upper Rhine

UPPER RHINE

Figure 1.   Project scale – local scale – river segment.

scale into account, inner city harbour conversions may therefore also indirectly contribute to flood mitigation. As our capacity to adapt to floods remains limited, all measures to contribute to the capacity to store water should be considered.

2   MAINZ ZOLLHAFEN IN CONTEXT

Mainz is located on the Upper Rhine downstream of the confluence with the river Main. Together with Frankfurt and Wiesbaden, Mainz is part of the Rhine-Main-Region with 4 million inhabitants. The Zollhafen Mainz is one of the largest container ports on the Upper Rhine with a transshipment of 1.6 million tons per year. Due to changing spatial demands, harbour activities are currently being moved to the site of the Industriehafen, directly adjoining the Zollhafen area to the northwest. In the former Zollhafen, an area of 22 hectares, the development of a new city quarter with 4,000 future jobs and 2,500 future inhabitants is planned. A masterplan defines the building layout and density (floor space index 2.8). The project site is situated in the flood plain where active flood defence measures are not allowed (EU Flood Directive 2007).

The Zollhafen is being developed as a model project for flood adapted building in a partnership between the Stadtwerke Mainz Corporation (the municipal utilities company) as the owner of the harbour and the Ministry for the Environment of the state of Rheinland-Pfalz. The partnership aims for the development of new prototypes for flood-resilient building and landscaping. Apart from meeting requirements for building in the flood plain, different strategies and instruments are being developed to trigger innovative adaptive design and to create public spaces which raise awareness for the inherent, but, due to the height of the site, rare flood risk.

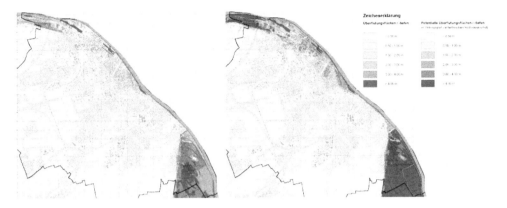

Figure 2.   Flood prone areas in Mainz for a 1/100–1/120 year flood (equivalent to 1/200 year flood when all retention polders along the Upper Rhine are installed) and for an extreme flood event exceeding the flood defense (source: Ministry for the Environment Rheinland-Pfalz).

Built in the flood plain to ensure the necessary water depths, harbour basins are often remnants of natural side channels. During the process of industrialization, the urban waterfront was expanded into the riverbed to channel the river, the basin as a concave excavation was left. This not only improved the navigability of the river, but also provided additional space for harbour activities. The plans for the Zollhafen development in Mainz projected on the landfill as part of the Upper Rhine rectification illustrates this development. What was formerly a gradual waterfront with wooden embankments now became an orthogonal, hard profile. It further implied a raise of the Zollhafen. This poses a challenge regarding flood risk awareness in the Zollhafen as it is only rarely flooded. The adjoining Neustadt lies behind the defense, but large parts were not raised. Therefore, the inherent flood risk in case of a breach or exceedance of the flood defense is much higher. The Rhine is the largest river in Western Europe. A 1/200 year flood implies a discharge of 8,000 m³/s at Mainz. The new flood hazard maps of Mainz show that more severe floods (between the 1/200 year event and the extreme flood of a 1/1000 year event, nearly 10.300 m³/s) will exceed the flood defences and expose 60,000 inhabitants of the city of Mainz to floods (Webler, 2010).

3   AWARENESS

The rare flood risk of the Zollhafen and the integration of the defense line for the adjoining city on site within the building plots demands for risk awareness of all involved. Flood risk awareness can be differentiated. There is awareness of approaching danger demanding for action, awareness of the adaptive capacities of our built environment (questions of materiality, etc.), an understanding of the environment on a more global scale regarding history and geography demanding for constant education, but also the awareness of the risk that something can go wrong (Redeker, 2010b). In the Zollhafen, flood risk awareness is not only incorporated spatially as a key aspect of the design strategy, but also in practice given the requirements to share operative responsibilities for developments outside of the flood defense.

4   SPATIAL PARAMETERS—OPEN SPACES AND BUILDING PLOTS

4.1   *Open spaces*

In a first step, retention-neutral development and safe access up until a 1/200 year flood was ensured via the height development of the open spaces. Streets and open spaces compensate

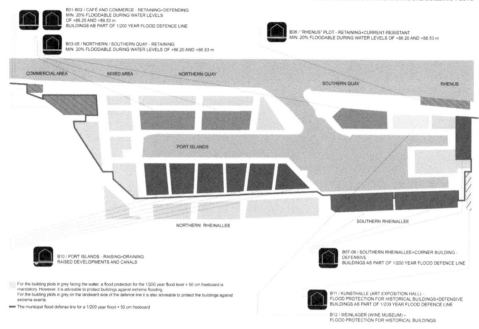

Figure 3.   Requirements for the individual building plots.

new developments fully. The difference between retention capacity, calculated as the volume defined vertically by the existing harbour morphology and the 1/100 year flood line and horizontally by the border to the river during average water levels and the border of the site defined by the existing defence. The Zollhafen has been designed to manage the risks should the development be threatened by flooding. The following scenarios have been addressed:1/100 year flood occurs – no threat, streets will not be flooded.

- 1/200 year flood occurs—streets will be partly flooded, but the main emergency routes will remain accessible.
- 1/200 year event plus 50 cm, the highest flood protection level for the City of Mainz - all streets in Zollhafen will be completely flooded, but dwellings remain unaffected.

Based on the developed geometry, the open spaces will be designed to convey the inherent dynamics in water levels. In preparation of a design competition for the open spaces, workshops took place with architects and urban designers and the municipal departments involved in operative flood risk management. It showed that spatial interventions to convey the flood risk during dry times did not always comply with operative flood management measures in case of an event. Once again taking the potential time scale between flood events into account, the general message is to keep solutions simple and sustainable. For example, the visibility of the evacuation routes could fall short due to power outings when designed with light. Another proposal for the defense line to function as a raised promenade, also during high water levels, would need to be well designed for flood tourists not to get in the way of operative measures.

Using mobile defense elements as urban furniture were among the ideas which were complemented. The elements could be used on site during average water levels. This solves the storage problem and enables direct access when the defense line needs to be completed. The redesign of the quay heads will be one of the main elements to raise public awareness for the flood risk of the Zollhafen by creating a public space that conveys historical, current and future water levels.

60

4.2 *Building plots*

At the border between the Zollhafen area and the Neustadt, a number of new buildings will accomodate the municipal defense line. On the street level businesses will be unaffected below the 1/100 year event. Dwellings are raised above 1.2 to 1.5 meters depending on their location. Where inundations of the site are not specifically foreseen, protection against higher floods will be provided by customized mobile defense systems for ground level programs, also for the two historical buildings on site, the Kunsthalle and the Weinlager. In addition, selected building plots will be designed to store water on site by accommodating at least 20–40% of the water between the 1/100 and the 1/200 flood. Only one building, today the site of the Rhenushalle 11, is located in the discharge area and directly affected by current and potential impact. Thus, it has to be developed in a way which does not negatively influence the discharge capacity and has to withstand potential impacts through floating refuse.

5 OPERATIVE MEASURES

The streets in the Zollhafen quarter will not be inundated until the flood level in the Rhine rises above 86.20 m (1/100 year flood). When the flood rises above 86.20 m (1/100) the rescue routes are progressively affected. Before the water level reaches 85,60 m the City of Mainz closes gaps in the flood defenses with temporary systems. This action alerts inhabitants that a flood is expected. When the water level reaches 85.95 m (1/75) the Authority for Civil Defence of Mainz initiates the municipal disaster management plan and the Civil Defence alerts inhabitants via loudspeaker or flyers. If the flood rises very quickly the Authority will also sound the warning siren.

The flood defence will be exceeded at a level of 87.00 m (1/200 plus 50 cm freeboard). Vast areas of the Mainz will be flooded. Even under these conditions the dwellings in the Zollhafen could still be inhabited. However, rescue vehicles will not be able to access the area. Also due to the rare flood frequency, uncertainty regarding the availability and type of rescue vehicles is a given. The Zollhafen may also be accessed by boats. The City of Mainz will decide whether evacuations have to be initiated and how. This decision will be influenced by forecasts regarding the height and duration of the flood. The City of Mainz is currently revising its emergency plan in accordance with the EU Flood Directive. Parameters for the evacuation of the Zollhafen and/or other city quarters will be defined and emergency routes, gathering points and provisional accommodation will be identified. The preparation of this plan is currently underway.

6 COMMUNICATION - PROJECT DEVELOPMENT GUIDE AND FLOOD RISK MANAGEMENT GUIDE

To communicate the requirements and also the design potentials of the concept for the individual building plots to project developers during the development phase and to provide information for the future inhabitants and users of the Zollhafen district, two handbooks were developed—the Project Development Guide and the Flood Risk Management Guide. According to the four A's of the FloodResilienCity Interreg IVb project, the PDG & FRMG aim to raise AWARENESS of the flood risk, to show how ALLEVIATION and AVOIDANCE may be factored into the development and to inform about the ASSISTANCE provided by the City of Mainz for the inhabitants.

6.1 *Project Development Guide (PDG)*

The PDG provides future developers with a framework for flood-resilient development of the individual sites. It explains the requirements that have to be met and the potential for innovative development that building in a flood-prone area provides. A legend defines spatial demands and possible solutions, such as defensive/mound+canal system/retention capacities on site/impact and current resistant. Solutions include external flood defence/integrated flood defence/reduced footprint/floodable. For each site, the PDG gives relevant excerpts from the development plan, a detailed description of requirements and specifications according to the position of the plot in the Zollhafen. Furthermore, a quantitative visualization of the requirements and possible solutions illustrated in sections (based on the design of the masterplan, buildings could also end up looking quite different) are provided.

Basic parameters for flood-adapted building are included in the legally binding layout plan currently underway. Beyond a building permit for the individual plots, building in the flood plain additionally requires a permit by water legislation. Building development in the flood plain was generally prohibited in Germany by federal law in 2005. Only very few exceptions are pemitted, one of them being the conversion of derelict inner city harbours. The PDG gives a step-by-step description of how to apply for a permit by water legislation. Further, reference projects are shown, organized according to the different possible solutions, for example:

- development on stilts where buildings cantilever onto the promenade, thus reducing the buildings footprint and ideally its damage potential (Hamburg)
- temporary flooding of underground car parks to avoid structural damage caused by buoyancy and to provide part of the volumetric compensation, the public part of the garage is separated from the flooded sector by steel gates and bulkheads (Cologne)
- hollow quay bodies, raised entrances and floodable garages (Frankfurt)
- defensive systems and their operational requirements
- integrated landscape design to create defensive functions (Almere)

Finally, a glossary for some of the crucial building elements and a list with flood-resilient materials, and their required qualitative properties is provided. This glossary is based on existing literature and provides links for further reading.

6.2 *Flood Risk Management Guide (FRMG)*

The FRMG adresses inhabitants, facility managers and other users of the Zollhafen. It explains responsibilities in case of a flood by giving answers to the 10 following questions:

- How high is the risk of being flooded?
- To what extent is my building protected?
- How can I protect my technical installations?
- How can I protect my car and movable equipment and protect the river against substance discharge?
- Can I insure myself against flood damage?
- When a flood is approaching: Where do I get information from?
- How will I be warned in case of a probable emergency?
- What does the local authority do to protect me?
- How do I have to prepare myself for an approaching flood?
- What do I have to do when the flood has arrived?

The FRMG closely follows the requirements of the EU Floods Directive 2007. It elaborates to what degree protection measures are integrated in the design of the open spaces and buildings and where responsibilities are in the hands of the inhabitants and facility managers. For example, by moving cars and other objects as some of the underground spaces will be flooded to avoid buoyancy. Also, furnishings should be water resistant or movable and materials that could be dangerous or contaminate flood water should be stored safely from floods. The guiding principle is to keep the damage potential low. The FRMG provides a number of check lists to assist people to prepare for and manage residual flood risks.

Regarding the insurance of buildings in the flood plain, flood resilient and resistant development makes the residual flood risk insurable. In Germany the so-called "elementary insurance" covers flood damage. But also here, the responsibility lies in the hands of the developers and inhabitants or building managers to adapt the buildings to the inherent risks. This includes the installation of temporal structures. In case of a flood information is provided via tv, radio and internet, telephone and cell phone. Flood forecasts can already provide trends 4 days in advance. This leaves due time for inhabitants to prepare and to take actions, given that they have been planned in advance.

7 CONCLUDING

How is the concept of resilience able to improve urban risk management?

People have at all times settled by the water. This must also be possible as flood risk increases. As a 19th century development, the Zollhafen was built into the raised flood plain. Consequently, all measures today to build flood-adaptedly have to consider the rare flood frequency to ensure adequate risk awareness and operative knowledge by all involved in case of an event. Our capacity to adapt remains limited as absolute protection against floods is not feasible. Although highly adapted due to a new risk awareness and according legislations, the residual risks rely on operative measures and knowledge by all involved. Therefore flood risk management plans for all possible scenarios is required (Theis, 2010).

To develop the Zollhafen as a flood resilient urban typology, a permanent integration of the flood risk into the design has been chosen. Specifically due to the height and safety level of the Zollhafen, the concept foresees inhabitants to manage on their own. Other parts of the city may be more affected due to their height, but also due to a lacking risk awareness behind the defense. The Zollhafen development is pushing for a new risk awareness beyond the project scale as questions being discussed for the site are in many cases interdependent with the adjoining city. Where defensive flood protection is not possible or acceptable, inundation must be allowed and dwellings must be adapted. This specifically applies for the existing building stock of areas behind the defence. Potential floods must be integrated in long-term redevelopment concepts.

REFERENCES

Bundesregierung 2002. *Perspektiven für Deutschland*. Berlin: Bundesregierung.
Bronstert, A. et al 2003. *LAHoR Quantifizierung des Einflusses der Landoberfläche und der Ausbau-maßnahmen am Gewässer auf die Hochwasserbedingungen im Rheingebiet*. Lelystad: CHR-KHR.
Haase, D., Nuissl, H. 2007. Does urban sprawl drive changes in the water balance and policy?: The case of Leipzig (Germany) 1870–2003. *Landscape and Urban Planning*, Volume 80, Issues 1–2: 1–13.
Meyer, H. 1999. Cities and Ports, Utrecht: International Books.
Theis, W. 2010. Five Guiding Principles for Flood Resilient Cities, Mainz: Ministry for the Environment, Forestry and Consumer Protection Rhineland-Palatinate, Department Water Resources Management (not published).
Redeker, C. 2010a. Zollhafen Mainz—Project Development Guide, Stadtwerke Mainz (not published).
Redeker, C. 2010b. Zollhafen Mainz—Reading the Floodplain, workshop documentation, Stadtwerke Mainz (not published).
Webler, H. 2010 Zollhafen Mainz—Flood Risk Management Guide, Stadtwerke Mainz (not published).

Resilience and Urban Risk Management – Serre, Barroca & Laganier (eds)
© 2013 Taylor & Francis Group, London, ISBN 978-0-415-62147-2

The Dutch room for the River programme and its European dimension

J. Tielen & J.M.T. Stam
Rijkswaterstaat—Room for the River, Utrecht, The Netherlands

ABSTRACT: This paper gives a brief overview of the Dutch Room for the River program. This program is currently being carried out to allow the Dutch river system to cope with higher peak flows. More than thirty measures are currently being carried out. One of these measures involves making a by-pass at he village of Lent, to protect the city of Nijmegen. This specific measure is also a part of a European project called FloodResilienCities. This EU project is very profitable for Room for the River and for the measure at Nijmegen-Lent because through the exchange with other cities new concepts are explored, new solutions for similar problems are studied and it can be seen how theoretical frameworks function in practice.

1   INTRODUCTION

1.1   *The plan*

The residents in the river regions of the Netherlands lived through anxious moments when in 1993 and 1995 the water levels were extremely high and the dikes just managed to hold. A quarter of a million people had to be evacuated. Extremely high river discharges are expected to occur more frequently in the future and for this reason it was decided to ensure that the rivers would be able to discharge the possible larger volumes of water without flooding. In 2007, the Central Government of The Netherlands approved the Room for the River Plan.

The Room for the River Plan has two objectives:

1. In 2015, the branches of the Rhine have to be able to cope with a discharge capacity of 16,000 cubic metres of water per second without flooding;
2. The measures implemented to increase safety will also improve the overall environmental quality of the river region.

The rivers in the Dutch delta are sometimes required to process huge amounts of water in very little time. In order to limit the risk of flooding, Rijkswaterstaat (the executive body of the Dutch Ministry of Infrastructure and Environment) is carrying out the 'Room for the River' program. A total of 17 partners—provinces, municipalities, water boards and Rijkswaterstaat are cooperating in the implementation of the Room for the River Program. The Minister of Infrastructure and the Environment has the overall responsibility for the Program.

Over the past centuries, the area available for the rivers has decreased continually. The rivers are confined between high dikes and an increasing number of people live behind these dikes. At the same time parts of the land behind the dikes has sunk due to soil subsidence. In addition, as it now rains harder and more frequently the rivers need to discharge more water to the sea. A flood in the current conditions would put the safety of 4 million people at risk.

The Netherlands is a country which traditionally is associated with water management. The inhabitants had to be inventive if they were to survive; thus they developed a highly sophisticated approach to water management (see Ven, 2000). The high population density (491 persons per square kilometre) adds pressure on available space and on the environment, which has to be managed carefully. Most of the economic activity is located in the low lying regions of the country. Since floods in these regions would cause damage in excess of 100 billion Euros, the country has adopted an approach of prevention rather than evacuation, resilience or reconstruction.

The continued strengthening of the dikes is an option that would reduce the risk of flooding. However, any flood that despite the protection, would occur, would result in even greater damage since more water would flood to the sunken land behind the dikes. This trend of continuously strengthening is changed to keep the Netherlands a safe, comfortable and pleasant country for its inhabitants. The answer lies in the 'Room for the River' plan. The Dutch government has taken the initiative to increase safety and to protect the land and people living behind the dikes from floods. The river will be given more room at over 30 locations covered by the 'Room for the River' Program. The main objectives of this program are to complete the flood protection measures around 2015 and to improve the overall environmental quality in the river region. In this way, a new approach to water management is implemented: instead of continuing to increase the height and size of the dikes, the Netherlands is now making more room for water.

1.2 *Types of measures*

The Room for the River program uses different solutions to give the rivers more room, some are explained below.

Lowering of flood plains: Lowering (excavating) an area of the floodplain increases the room for the river at high water levels.

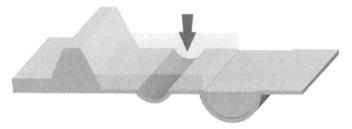

Dike relocation: Relocating a dike land inwards increases the width of the floodplains and so provides more room for the river.

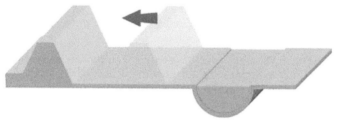

Lowering groynes: Groynes stabilise the location of the river and ensure that the river remains at the correct depth. However, at high water levels groynes can form an obstruction to the flow of water in the river. Lowering groynes increases the flow rate of the water.

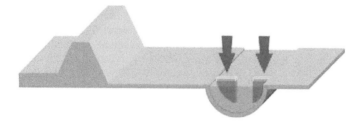

High-water channel: A high-water channel is a diked area that branches off from the main river to discharge some of the water via a separate route in high water circumstances.

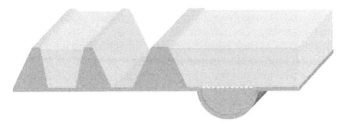

Depoldering: The dike on the river side of a polder is relocated land inwards. The polder is depoldered and water can flood the area at high water levels.
Removing obstacles

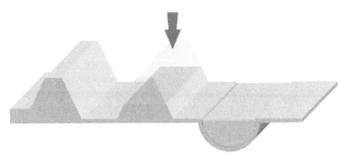

Removing or modifying obstacles in the river bed where possible, or modifying them, increases the flow rate of the water in the river.

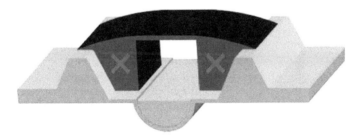

As a last possibility dikes are strengthened in areas in which creating more room for the river is not an option.

2 PROGRESS OF THE PROGRAM

2.1 *General progress*

The program started in 2007 and should be finished in 2015. In 2011, half way through the program, a mid-term review has been made to investigate whether the objectives are being obtained. Some conclusions of the mid-term review were as follows.

To keep up the pace it has been very important to make clear from the start that the budget for the projects was available: a budget of €2.2 billion has been released since 2007. This avoids budget discussions during the program; when a project matches the conditions that were stated in the Key Decision, its progress will continue step by step. Also a clear deadline is very important to stimulate all parties involved to avoid hesitations, long decision times and to organize activities in parallel.

From the start the central government has promoted cooperation with local authorities; the centrally accorded plan should be executed, as much as possible, by regional or local authorities to ensure local support and to use local knowledge of the people in the area. In this way many of the projects were adopted by provinces, water boards or municipalities.

To keep control on the time ad budget of the programme, the central government defined Room for the River as a "Large Project" which has to report twice a year on developments, scope, budget and planning. The parliament is informed on these issues with a progress report, which they discuss in their regular consultation gathering with the secretary of state. This political attention is another incentive for good progress.

The past four years the main part of the measures were further developed in design plans and permits. At the Overdiepse Polder and some five other projects, the building and excavation activities have already started. In 2012 procedures for permits and spatial plans will be followed and brought to an end for many other projects, so that they will be able to start the execution of the works. For at least four of the projects it will be very difficult to achieve the dead line of end 2015.

2.2 *Changes in people's awareness*

Like in many other countries where people are well protected from flooding, there is a short memory on the constant need for measures. People feel safe behind the dikes. They expect their government to protect them thoroughly and this policy of avoidance is mainstream for the Dutch. And because this policy has been quite successful over the last centuries, the public doesn't see the need for other ways of handling high waters. Acceptance of any inconvenience or damage is low and therefore discussions on adaptation or resilience are only just beginning.

After some 15 years with no serious threats the awareness about the dangers of flooding decreases quickly and the sense of urgency for the measures is endangered.

Questions that are posed are: Are all those measures really necessary? Are the standards for safety perhaps too high? Are we really at risk? Why doesn't Germany solve this problem for us? Half way the program there even has been a fundamental discussion in parliament about the need for all projects along the river IJssel (the most northern branch of the Rhine).

These questions also come up in councils of the municipalities involved. The responsible authorities sometimes hesitate to take decisions (e.g. on spatial plans) because citizens are severely affected by the projects and search for support from their representatives. This results in delays and longer periods of uncertainty for the people involved. One of the causes for these hesitations is the long stretch of time involved for the planning process. For the average project it takes 8–10 years from start to finish and during that time elections on all levels result in new politicians with new issues causing political attention for the Room for the River program to vary strongly.

As a proverb from the water boards says (in Dutch it rimes): "Give us this day our daily bread and every year a water flood", the public needs to remain aware of the dangers of flooding to keep support for the more than 30 projects in the program.

2.3 *An example: Dyke relocation in Nijmegen*

One of the Room for the River measures is the Room for the river Waal in Nijmegen. The river Waal has a sharp bend near Nijmegen and forms a narrow bottleneck for the flow. In the years 1993 and 1995 this caused near floods. In the case of Nijmegen, the Waal dike in Lent will be relocated and a new channel in the newly created flood plains will be constructed. This will create an island in the Waal and a unique urban river park with lots of possibilities for recreation, culture, water and nature. The solution is far-reaching, yet sustainable and safe. The actual digging is to start in 2013. The relocation of the dike, the construction of the new channel and the island is scheduled for 2016. The area will be further developed in the years after that to allow for recreation, housing facilities and other urban functions.

In short the measures to be taken are: the dike is to be moved 350 m inland. A new channel is to be dug in order to give the river more room. This will create an elongated island. Bridges across the ancillary channel have to be build to guarantee the accessibility of all parts of the city.

3 THE EUROPEAN DIMENSION OF ROOM FOR THE RIVER

3.1 *The FloodResilienCity project*

The Room for the River program benefits from cooperation and knowledge exchange with other European neighbours that cope with similar problems. One of these cooperations is the INTERREG IVB NWE project FloodResilienCity (FRC). FRC is focussed on flood management in urban areas. It is a cooperation between the public authorities of eight European cities and two universities aimed at improving flood management in urban areas. The Room for the River measure at Nijmegen is one of these partners in FRC.

The motivation for initiating this cooperation lies in a shared desire to integrate the increasing demand for more houses and other buildings in urban areas with the increasing need for more and better flood risk management measures in North West European cities along rivers. Demographic resettlement and increasing populations in North West European countries have resulted in pressure on cities to build more houses. Furthermore commerce and industry constantly increase the need for building in and around cities. Conflicts can arise between urbanisation and flood risks in particular in cities. The FloodResilienCity project wants to turn the problem into a positive opportunity for the further development of cities in North West Europe: More room for city and water resulting in multiple benefits and a more attractive city to work and live in.

3.2 *A wide strategy to flood management*

The FloodResilienCity project has used an approach, termed as the four A's approach. The four A's cover the spectrum of approaches of the safety chain. They are awareness, alleviation, avoidance and assistance.

– Awareness means increasing the consciousness of flood risks and what can be done about it. Rivers have been canalized and sometimes completely culveted so that people scarcely can remember that a river runs under their city (as in Bradford). Redesigning public spaces so that people in cities become aware of the river helps. For example in Orleans (France) the new quay walls and banks are designed so that people can recreate and enjoy the river.
– Avoidance includes all activities that limit flood damage and ease recovery, for example flood proofing buildings and infrastructure. City quarters close to the river can be designed to withstand a higher flooding frequency. For example the old harbour area in Mainz, which is being revitalized to become a flood proof residential and business quarter.
– Alleviation involves reducing flood risk by implementing physical, technical and procedural measures. Dykes and dams belong to this category but another interesting concept is using streets as streams – which means adapting streets so that surface run-off can flow safely without flooding adjacent buildings (as is being done in Dublin – Ireland). Flood reservoirs and overflow areas are also typical alleviation measures. They are constructed to protect major cities such as Paris or Brussels from flooding. The neighbouring rural areas that often have to provide the space for these measures—are thus confronted with flood protection ensuing discussions with stakeholders about compensation measures.
– Assistance implies support for recovery and capacity building in communities. Emergency routes and plans and strengthening the organizational infrastructure are typical alleviation measures. Emergency plans, specially in urban areas, will be very complex including problems as how to maintain the cities services (waste management, social services etc.) in times of floods.

The FloodResilienCity project will ultimately result in better solutions, more awareness and increased capacity in flood management in the cities of Bradford, Brussels, Dublin, Leuven, Mainz, Nijmegen, Orléans and Paris. The project will activate a structural change in the mindset of the politicians, professionals and public in these partner cities. That change concerns an integrated approach in their sustainable flood risk management policies. The main output of the FRC project is an intense cooperation between 11 partners in 8 cities in

North West Europe. The regional investments deliver tangible results in each partner city. The transnational actions will prepare the ground to implement the FloodResilienCity strategy in plans and policies of each partner city.

3.3 *Lessons for Room for the River*

The Room for the River programme has a broad goal towards the prevention of flooding in the Netherlands. The main input for the FloodResilienCity project from Room for the River Nijmegen is the spatial design and use of the new island and of the waterfront on the North Quay of Lent (Nijmegen). Other issues such as the technical design of the measure against seepage, the design of the emergency routes to and from a newly created island and the communication plan for all different kinds of inhabitants of the redeveloped area in Lent (Nijmegen) are also part of FloodResilienCity.

There are three types of 'lessons' that are specially relevant for Room for the River and specifically the project of Lent.

– Applying each other's concepts: the Conseil de l'Agglomeration d'Orleans wanted to raise the awareness to flooding by redesigning the public space next to the Loire. They removed car parks and opened the area so that people could walk and recreate next to the river. The quays were restored with soft slopes in a way that the public can reach the river. The objective was to get the inhabitants starting to live with their faces towards the river instead of with their back to the river. This concept, gently sloping quay walls to tempt people approaching the river, was adopted by the city of Nijmegen in their design of the island that will remain in the river after the new channel has been made.
– New contexts that result in 'eye-openers': In the Netherlands as a general rule, building in the flood plain is forbidden, as the level of safety prescribed by law cannot be guaranteed. This is very different to the case of Mainz, where the Stadtwerke Mainz is redeveloping the old customs harbour area into a zone which combines shopping, living (2500 inhabitants) and working (4000 jobs). The Zollhafen quarter will have a higher risk of flooding than the rest of Mainz, but will be designed for this situation (i.e. with a probability of flooding of 1:200 years, the streets will be partly flooded).
– Comparing theory and practice: several partners have been working at contingency plans as part of the assistance approach. The Conseil-Général du Loiret has developed a master plan for emergencies for which also an exercise was held. Other partners are making evacuation plans also for when the water surpasses a designed level and an expected evacuation in a designated area has to take place (e.g. in Zollhafen Mainz or in the future island of lent in Nijmegen). Comparing these plans to life experience of unexpected emergency flooding such as in Flanders (November 2010) and (Dublin October 2011) helps to test them for robustness and effectiveness.

4 CONCLUSIONS

The Dutch Room for the River program proved to be a change in the traditional policy of strengthening dykes. It was initiated after a major flood for which a quarter of a million people had to be evacuated. The program consists of more than thirty measures. AS of now, the program is still on schedule. The major risk is the changes in people's perception and sense of urgency which could constitute a risk for the political commitment to the program.

The program profits from exchanges with other European organisations with similar problems. One example is the INTERREG NEW IVB project FloodResilienCity. The first important lesson is that flood management can be approache din the broad strategy of the safety chain. In FloodResilienCity this is termed with the four A's: Awareness, Avoidance, Alleviation and Assistance. The participating cities have different solutions within each of these four A's and this has led to new insights for Room for the River and for the project of Nijmegen-Lent.

ACKNOWLEDGEMENTS

The work reported in this paper was carried out with financial support from the European Union (INTERREG. NEW IVB project FloodResilienCity).

REFERENCES

Anonymous 2006. Spatial planning key decision, room for the river. Approved decision 19 December 2006.

Ven, G.P. van de (ed.), Man-made Lowlands. History of water management and land reclamation in The Netherlands (Utrecht: Matrijs, 2000).

Resilience and Urban Risk Management – Serre, Barroca & Laganier (eds)
© *2013 Taylor & Francis Group, London, ISBN 978-0-415-62147-2*

Managing urban flooding in the face of continuous change[1]

Chris Zevenbergen
UNESCO-IHE Delft, The Netherlands
TuDelft, Delft, The Netherlands

Assela Pathirana
UNESCO-IHE Delft, The Netherlands

ABSTRACT: A major cause of increasing flood risk of cities is the increasing population and value of domestic and commercial buildings and growing interdependency on infrastructure networks. Contemporary flood risk analyses generally include only direct economic damages and fail to consider other flood risks, such as disruption to the economy or the loss of life, which may be mitigated by building resilience into high value assets, critical infrastructure systems and urban communities. The infrastructure networks are critical for the continuity of economic activities as well as for the people's basic living needs. Resilience in cities depends both on its physical form and characteristics as well as on the people's capacity, and social behavior. There is a growing recognition that innovative planning approaches and processes based on these resilience principles will guide citizens and other stakeholders the way to become co-producers of a sustainable community that can respond to change and disruption, and pro-actively reduce vulnerabilities. In most industrialized countries, the building stock and infrastructure are mainly ageing and there is much heritage. In the coming decades, the redevelopment (c.f. renovation and modernization) of the existing stock is a high priority and certainly of higher priority than the provision of new housing. Redevelopment projects may thus provide windows of opportunity to make adjustments in the process of urban renewal in order to correct old mistakes and to build resilience by adapting and restructuring the urban fabric to new conditions of increased flood risk. Some cities in the developing world, however, is not often constrained by significant past investments, and much of the change in their urban fabric is to come in the next few decades. There is a huge challenge to exploit this momentum. If we are able to seize these windows of opportunity and share good practices via city-to-city networks stretching across country boundaries and other social networks, than we can create the groundswell for real practical change towards flood-resilient cities at a global level. This can convert the challenge into an opportunity.

1 INTRODUCTION

Many cities around the world are facing the challenges of sustainable living and development and are exploring ways to enhance their ability to manage an uncertain future. Drivers and pressures include relative wealth; population growth; the provision of food; lifestyle expectations; energy and resource use and climate change. These pose new challenges for the way in which we manage urban floods. There is no clear cut, 'best' solution for the avoidance of catastrophic flood events or even how to 'live with (all) floods' (Milly *et al.*, 2008). The way forward is thus far from clear although what we can be sure about is that we are rapidly entering a phase of fundamental change and our willingness and ability to adapt to and mitigate the worst effects of this will be critical (Pitt, 2008; Evans et al., 2005).

---

1. Adapted from Zevenbergen *et al.* 2010.

We live in 'yesterday's' cities. Many of the urban patterns that we see today—such as city layouts, buildings, roads and land ownership—are legacies of up to a century and a half of urban policy and decision-making; even longer in some of our cities. Tomorrow's cities will also be shaped by the decisions we make today. They must respond to more rapid changes in physical, social, economic and institutional conditions than recent generations have been used to.

In general, cities are becoming larger and denser. Urban expansion is an issue of serious concern and is often placed as a justification for densification. The fundamental question of whether urban expansion should be resisted, accepted or welcomed is still largely unresolved. From the perspective of flooding, concerns for indiscriminate urban expansion or 'sprawl' have captured the attention of both policymakers and academics during the last decade. This is because, alongside climate change, it is considered as the major driver for increased flood risk. Sprawl will occur where unplanned, decentralized development dominates, as is common in developing countries. Where growth around the periphery of the city is coordinated by a strong urban policy, more compact and less vulnerable forms of urban development can be secured. It is evident that these approaches to development have direct consequences for the way floods are managed both in terms of the vulnerability of the urban area and its inhabitants and also in terms of the often indiscriminate effect that urban growth has on the generation of floods in terms of runoff and flood probabilities.

2   URBANISATION

Urbanization, both as a social phenomenon and physical transformation, is driven by processes that take place at varying temporal scales from relatively slow (e.g. migration, rising water demand, sea level rise and changes in laws) to rapid (e.g. natural disasters, changes in regulations and economic systems). While there is much that is uncertain about the urban future, some recent experiences show that some urbanization pathways are more desirable than others because they will likely lead to more (flood) resilient cities. Population development is often dictated by factors other than flood vulnerability: currently in cities like Tokyo, Yango and Dhaka in Asia, rapid urbanization into flood prone areas are happening (Adikari, 2010). Cities can grow with infill (case with many megacities in the third world) and expansion (e.g. suburbanization as was common in the urban development of united states) resulting in dramatic changes in level of resilience. Even with infill driven densification, exploiting urban regeneration opportunities and therefore influencing the urbanization pathways, can help improving flood resilience. For example the urban redevelopment of some neighborhoods of Dordrecht city, Netherlands. (Gersonius et al. 2012). Some cities a shrinking in size. While depopulation of cities does not by itself reduce the flood risk, opportunities can be exploited to increase the flood resilience of the cities while improving overall appeal of the cities by means like introducing green infrastructure (Schilling and Logan, 2008).

These experiences highlight the need to take a completely new and different perspective on urban design, planning, and building. Creative thinking and innovations in socio-economic and technological systems are essential to change existing management structures and regimes. There is a growing recognition that responses that enhance resilience, can be implemented gradually in combination with autonomous retrofitting, and offer prospects for action in the short term in regional planning and development in cities. These interventions should operate in a mode of constant learning and experimentation and follow an adaptive, cyclic approach rather than a linear and predictive one. Such interventions do not only reduce flood impacts, but also create new opportunities and co-benefits.

3   VULNERABILITY OF CITIES

A major cause of increasing flood risk of cities is the increasing population and value of domestic and commercial buildings and growing interdependency on infrastructure networks (i.e. power supplies, communications, water, transport etc.). The infrastructure networks are critical for the continuity of economic activities as well as for the people's basic living needs. Their availability is also required for fast and effective recovery after flood disasters. The severity of flood damage therefore is often largely dependent on the degree that both high value assets and critical infrastructure systems are affected, either directly or indirectly (Serre *et al.*, 2011). Whilst ageing infrastructure and building stock in the developed world pose a risk due to increasing vulnerability, this also provides an opportunity to introduce new technologies in the redevelopment process and to adapt infrastructure and buildings to enhance flood resilience (Gersonius *et al.*, 2010)). Urban restoration, regeneration and modernization can be a key driver of economic development, both as a result of the initial investments required and the benefits that will accrue over time (e.g. formerly flood-prone areas may become available for productive use).

In 2008 the OECD has made a first estimate of the exposure of the world's large port cities (n = 136) to coastal flooding due to storm surge. It also included estimates of the impact of future climate change (Nichols *et al.*, 2008) (see Figure 1). The study revealed that a large number of people living in mega-cities today are already exposed to coastal flooding: about 40 million people in large port cities are exposed to a 1 in 100 year coastal flood event. The exposure is concentrated in a few cities of which the majority are located in deltas. The cities with the highest population exposure today are almost equally split between developed and developing countries. Because the wealth of cities is higher in developed countries the same holds true for the total value of assets exposed to coastal flooding. Based on these results

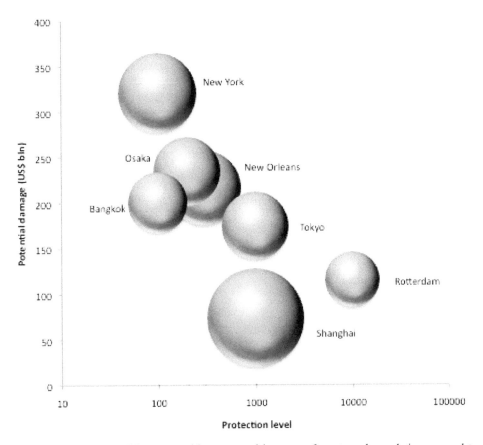

Figure 1. Vulnerability of large port cities expressed in terms of assets and population exposed to coastal flooding as a function of the protection level (based on data from Nicholls *et al.* (2008)).

they conclude that large city flooding may remain a frequent event at the global scale because so many cities are threatened (there is a 74% chance of having one or more of the 136 cities affected by a 100-year event every year assuming that flooding events are independent). Finally, the study has substantiated earlier analysis that show that socio-economic changes are the most important driver of the overall increase in both population and asset exposure. Climate change has the potential to exacerbate these effects.

4 RESILIENCE

The term 'resilience' is often used in discussions about sustainability. Sometimes, resilience is a more useful concept than sustainability, for instance when it is used within the context of sustainable urbanization. This is partly because resilience embraces explicitly the dynamic nature of (complex) systems such as cities, whereas sustainability is often conceived as a goal to which we should collectively aspire. For others, however, sustainability is an attribute of dynamic, adaptive systems that are able to flourish and grow in the face of change. Resilience- focused strategies promote the capacity of a system to cope with and recover from disturbance and to reorganize while undergoing change as core features of the system (Gersonius *et al.*, 2011).

Resilience in cities depends both on its physical form and characteristics as well as on the people's capacity, and social behavior. Community resilience requires self-reliant, skilled and capable citizens who have 'developed iterative learning with mature peer-to-peer social networks'. There is a growing recognition that innovative planning approaches and processes based on these resilience principles will guide citizens and other stakeholders the way to become co-producers of a sustainable community that can respond to change and disruption, and pro-actively reduce vulnerabilities. However, the approaches (and processes) for achieving this dynamism should not be viewed as models that can be applied in all contexts since they are shaped by the social and cultural norms of particular locality.

There is no single 'magic' recipe for successful planning of a city in response to the challenges of sustainability, climate change and flood risks. This is partly because every city has a unique context. What we have learned is that urban design, master planning and the management of buildings, infrastructure, public utilities and green infrastructure must be included in any urban flood-risk management strategy. We also learned of the need for long-term planning. A long-term perspective allows us to identify opportunities for synergy and to overcome barriers for implementation, such as investments that both enhance resilience and provide short-term additional economic, social or environmental benefits. A long-term perspective is also fundamental for incorporating sustainability indicators, such as life cycle cost. Planning with a long-term perspective thus opens the way to develop strategies that are more resilient, adaptable and responsive. It also requires skilled and capable stakeholders who are knowledgeable about the systems they live in and are capable of mainstreaming flood-risk management in the process of (re)development.

5 EPILOGUE

In most industrialized countries, the building stock and infrastructure are mainly ageing and there is much heritage. In the coming decades, the redevelopment (c.f. renovation and modernization) of the existing stock is a high priority and certainly of higher priority than the provision of new housing. European cities are composed of mixtures of buildings of different ages and life spans, but within 30 years, around one-third of its building stock will probably be renewed (ECTP; 2005). The same holds true for many other cities of the Western world, where continuous restructuring will be common practice. Redevelopment projects may thus provide windows of opportunity to make adjustments in the process of urban renewal in order to correct old mistakes and to build resilience by adapting and restructuring the urban fabric to new conditions of increased flood risk. Some cities in the developing world, however, is not often constrained by significant past investments, and much of the change in their urban fabric is to come in the next few decades. There is a huge challenge to exploit this momentum. If we are able to seize these windows of opportunity and share good practices via city-to-city networks stretching across country boundaries and other social networks, than we can create the groundswell for real practical change towards flood-resilient cities at a global level. This can convert the challenge into an opportunity. There are a growing number of emerging examples of innovatory initiatives changing the way in which these challenges are being addressed and of which we can learn! (Birkmann *et al.*, 2010).

REFERENCES

Adikari, Y., Osti, R., & Noro, T. (2010). Flood-related disaster vulnerability: an impending crisis of megacities in Asia. *Journal of Flood Risk Management*, 3, 185–191. Retrieved from http://onlinelibrary.wiley.com/doi/10.1111/j.1753–318X.2010.01068.x/full

Birkmann J., Garschagen M., Kraas F., Quang N. (2010). Adaptive urban governance: new challenges for the second generation of urban adaptation strategies to climate change. Sustain Sci. (2010) 5:185–206 DOI 10.1007/s11625–010–0111–3.

ECTP (2005) Strategic Research Agenda for the European Construction Sector. European construction technology platform Onine [http://cordis.europa.eu/technology-platforms/pdf/ectp2.pdf] accessed May 2010.

Evans E.P., Ashley, R. M., Hall, J., Penning-Rowsell, E., Saul, A., Sayers, P., Thorne, C., Watkinson, A. (2004). Foresight. Future Flooding Vol I – Future risks and their drivers. Office of Science and Technology. April. Evans, E.P., Simm, J.D., Thorne, C.R., Arnell, N.W., Ashley, R.M., Hess, T.M., Lane, S.N., Morris, J., Nicholls, R.J., Penning-Rowsell, E.C., Reynard, N.S., Saul, A.J., Tapsell, S.M., Watkinson, A.R. and Wheater, H.S. (2008) An update of the Foresight Future Flooding 2004 qualitative risk analysis. Cabinet Office, London.

Gersonius, B., Nasruddin, F., & Ashley, R. (2012). Developing the evidence base for mainstreaming adaptation of stormwater systems to climate change. *Water Research*. Retrieved from http://www.sciencedirect.com/science/article/pii/S0043135412002345

Gersonius B., Ashley R., Pathirana A. and Zevenbergen C. (2010). An options planning and analysis process for managing the flooding system's resiliency to climate change. Proc. Of the Institution of Civil Engineers. Engineering Sustainability. March Vol. 163, Issue E1, p15–22. Doi: 10.1680. ensu.2010.163.1.15. Paper 900020.

Nichols, R.J., S. Hanson, C. Herweijer, N. Patmore, S. Hallegate, J. Corfee-Morlot, J. Chateau and R.M. Wood. Ranking Port Cities with high exposure and vulnerability to climate extremes: exposure estimates. OECD EnvironmentalWorking papers No. 1, OECD Publishing.

Milly, P.C.D., Betancourt, J., Falkenmark, M., Hirsch, R.M., Kundzewicz, Z.W., Lettenmaier, D.P. and Stouffer, R.J.(2008). Stationarity Is Dead: Whither Water Management? Science, 319 (5863), 573–574.

Pitt M (2008). Learning Lessons from the 2007 Floods. HMSO.

Schilling, J., & Logan, J. (2008). Greening the Rust Belt: A Green Infrastructure Model for Right Sizing America's Shrinking Cities. *Journal of the American Planning Association*, 74(4), 451–466. Retrieved from http://www.esf.edu/cue/documents/Greeningtherustbelt.pdf

Serre, D., Lhomme, S., Heilemann, K.,Hafskjod, L.S., Tagg, A. Walliman, N. and Diab, Y. (2011) Assessing vulnerability to floods of the built environment—Integrating urban networks and buildings. Proceedings ICVRAM 2011 and ISUMA 2011, 746–753.

Zevenbergen, C., Veerbeek, W., Gersonius, B and van Herk, S. (2008) Challenges in urban flood management: travelling across spatial and temporal scales. J Flood Risk Management, 1:2(81–88).

Zevenbergen, C. Cashman, A. Evelpidou, N. Pasche, E. Garvin, S. and Ashley, R. (2010) Urban Flood *Management, Tailor & Francis* ISBN-10: 0415559448 | ISBN-13: 978–0415559447 |

Resilience and Urban Risk Management – Serre, Barroca & Laganier (eds)
© 2013 Taylor & Francis Group, London, ISBN 978-0-415-62147-2

Urban resilience in post-Katrina/Rita New Orleans

J.R. Amdal Sr.
Merritt C. Becker Jr. University of New Orleans Transportation Institute, New Orleans, LA, US

ABSTRACT: From the time of its founding in 1718, New Orleans incorporated resiliency by building raised structures on naturally high ground, separating living space from potential floodwaters. Residents understood and respected the natural limitations of the city. However, in the late 1900s, with the introduction of drainage pumps, outfall canals and equipment stations, New Orleans envisioned a city no longer constrained by either geography or tradition. Below-sea-level areas were drained and became footprints for new neighborhoods. After 1965's Hurricane Betsy, the Corps of Engineers undertook a series of construction projects throughout the New Orleans region to protect from future hurricanes. In 2005, Hurricane Katrina proved the city's storm protection system unfit to handle natural disasters, when eighty percent of New Orleans flooded. Consequently, government leaders, citizens, professionals and academic advisors have worked to achieve enhanced urban resiliency through recovery projects. New Orleans has embarked on its post-disaster journey by rethinking and reimagining the city and its neighborhoods for the betterment of all its citizens in an era of strengthening hurricanes, rising sea levels and climate change.

1 INTRODUCTION

Pierce Lewis, a world-renowned geographer and chronicler of New Orleans, famously described this place as "The Impossible but Inevitable City" (Lewis, 1973). Since its founding in 1718, New Orleans has occupied a perilous but strategic location for the nation at the base of the Mississippi River system. Low in elevation, flood-prone, swampy, water-logged, insect-ridden, New Orleans has been subjected to many disasters over the last 300 years: tropical diseases, torrential rains, annual floods and catastrophic hurricanes. However, the worst to date were Hurricanes Katrina and Rita, which struck the Gulf Coast and New Orleans in August and September 2005, respectively.

These storms reinforced New Orleans' perilous position in the eyes of the nation and the world. But since that fateful day, August 29, 2005, New Orleans has demonstrated its resiliency as a city, reaffirmed the people's belief in its future and cemented the bond its citizens have with this unique city. Today, the city is back after overcoming seemingly insurmountable odds. The story of New Orleans' post-Katrina/Rita recovery is a long and drawn out tale of trial, tribulation and, ultimately, triumph but it remains a work in progress.

Hurricane Katrina emerged on the national and regional weather alerts in the last days of August 2005 as a monster storm affecting the entire Gulf of Mexico and heading directly towards New Orleans. With landfall expected during the early morning hours of August 29, Mayor Ray Nagin called for an emergency evacuation of New Orleans at 9:30 AM on August 28, 2005. In the hours before touching coastal Louisiana some 60 miles south of New Orleans, evacuation of the city and its surrounding parishes proceeded with some urgency, but no one expected the normal three-day hurricane inconvenience to stretch into weeks and, for some residents, months or even years.

Katrina, for New Orleans, was both a natural and man-made disaster. The storm's winds and waters took their toll, but the actual devastation in the city resulted from the failure of its primary flood protection systems: a series of earthen levees, concrete floodwalls and massive drainage canals. When these failed, 80 percent of New Orleans flooded.

Urban resiliency has had a unique history in the traditional growth and development of New Orleans. Our forefathers built appropriate structures with raised first floors on "high" ground. However these traditions were less common at the end of the 19th Century when new mechanical devices allowed the city to drain large portions of former swamplands and greatly enlarge the city's developable area. However, as Hurricane Katrina tragically demonstrated in 2005, "man and machine" can only do so much to withstand the forces of Mother Nature. During Hurricane Katrina and shortly thereafter with Hurricane Rita, our man-made fortifications against water and wind failed, resulting in the costliest disaster in the history of the United States. In the ensuing years, countless improvements have been made to our flood protection systems, our evacuation planning processes, and specific rebuilding programs to reintroduce resiliency into New Orleans' recovery.

2 TRADITIONAL NEW ORLEANS: "CITY BUILDING THAT RESPECTED NATURE"

Given New Orleans' unique geography (on a deltaic plain next to the Mississippi River in a flood-prone area with low elevations), citizens originally built raised buildings on high ground in anticipation of annual floods. Low-lying areas of the city remained undeveloped until the later part of the 1800s. Our forefathers understood urban resiliency: they built appropriate buildings (raised structures) in appropriate places (high ground).

Soon after the founding of New Orleans in 1718, owners of upriver plantations, citizens of small river towns, and residents of the city started building earthen levees along the banks of the Mississippi River to protect themselves and their property from its annual floods. These earthen levees offered little protection from hurricanes and were frequently overtopped or breached during periods of high water, but they did offer some protection under normal river conditions. These primitive forms of flood protection, although strengthened and raised over the years, still retain their vital function today.

Figure 1.   Historic french quarter streetscape. Photo courtesy of James Amdal, 2012.

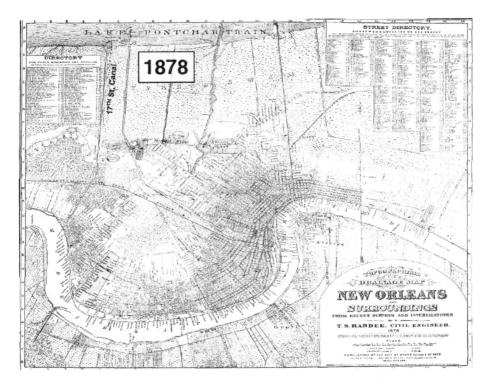

Figure 2.   Civil Engineer T.S. Hardee's map of the City of New Orleans, 1878. Photo courtesy of Louisiana State Museum.

In large part, early settlers had the right idea. They built where the city was most resilient, on naturally high ground. An 1878 map by Civil Engineer T.S. Hardee shows the City of New Orleans with 200,000 residents confined to the "Sliver by the River," an area of the city adjoining the river that suffered only minor flooding from Hurricane Katrina. Areas labeled "cypress swamp" on this and earlier maps include present day neighborhoods including Lakeview, most of Gentilly, New Orleans East, Broadmoor and the Lower 9th Ward. During Katrina, these neighborhoods received the worst flooding thanks to their naturally low elevations. It is notable that the unflooded areas on the post-Katrina map are almost identical to the populated areas of New Orleans on the 1878 map. Our historic neighborhoods and their flood-conscious architecture weathered the storm with minimal damage and demonstrated that traditional New Orleans' settlement patterns and construction techniques resulted in resiliency. Neighborhoods within the "Sliver by the River" were the first to repopulate and today have the highest real estate values. In pre-Katrina New Orleans, the real estate maxim was "Location-Location-Location". Post-Katrina it became "Elevation-Elevation-Elevation."

3   A NEW CITY-BUILDING CONCEPT FOR NEW ORLEANS: "MAN AND MACHINE BEATS NATURE"

Starting in the late 1890s, New Orleans began to change in a rather remarkable manner. The concept of "Man and Machine Beats Nature" was introduced by the city's Sewerage and Water Board (SWB) when they began constructing a world-renowned drainage system. Using the massive Woods Screw Pumps—invented by a New Orleans engineer in the early 1900s—as well as pump stations and outfall canals, the SWB drained the city's low lands. This massive construction program dramatically increased the city's buildable area by making vast areas of the city developable. This was the first large-scale attempt to beat Mother Nature in New Orleans. It would not be the last.

After Hurricane Betsy devastated New Orleans as a Category 4 storm in 1965, the US Army Corps of Engineers was directed by Congress to build an extensive flood protection system around the city. This system was comprised of higher and stronger earthen levees and massive concrete floodwalls. Unfortunately, this flood protection system was not complete after 40 years of construction when Katrina hit New Orleans in 2005. It remains incomplete today, although it is nearing completion. With these manmade interventions, residents in New Orleans assumed they were safe and secure. They truly believed they lived in a resilient city. Unfortunately, the flaws of their flood protection system were tragically exposed by Katrina's wrath.

4   HURRICANES KATRINA AND RITA: A DOUBLE WHAMMY

Hurricane Katrina was the most destructive as well as the costliest natural and manmade disaster in the history of the US. Katrina was a massive Category 3 storm with winds in excess of 125 mph. This hurricane impacted the entire Central Gulf Coast including the states of Louisiana, Mississippi, Alabama and Florida; not just New Orleans and southern Louisiana. While the Mississippi Gulf Coast was especially hard hit by Katrina's winds and 30-foot-plus storm surges, 80 percent of New Orleans flooded from breaches in our federally designed and constructed flood protection system. All utilities and telecommunications systems failed. All surface transportation systems were flooded. Many roadways were impassable for months while some, such as the Twin Spans Bridge over Lake Pontchartrain, were completely destroyed. Basic utility services as well as communications (cellular and landline) remained severely compromised for months. The Port of New Orleans was an anomaly. It reopened after just 13 days, thanks to access provided by the Mississippi River and a cleared navigational channel provided by both public and private sector forces.

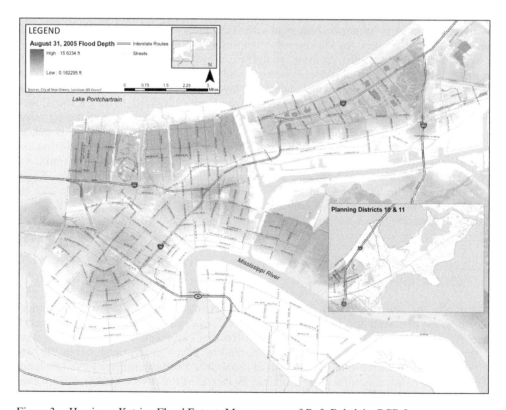

Figure 3.   Hurricane Katrina Flood Extent. Map courtesy of Rafe Rabalais, GCR Inc.

For New Orleans, Katrina was a natural and a manmade disaster, one whose effects were made worse by our natural topography and the man-made protection systems that failed. New Orleans is naturally shaped like a bowl with its outer edges formed by earthen levees, floodwalls or natural ridges. When the manmade structures failed during Katrina, the "bowl" filled with water. In a weird turn of fate, the very structures that were meant to keep the water out of the city kept the floodwaters in the city until pumps were able to "dewater" New Orleans. This took 53 days. As has often been reported, roughly 50 percent of the Greater New Orleans region lies below sea level, which compounded this effort.

The dark arrows on the map below, prepared by the staff of the Times-Picayune, New Orleans' local newspaper, indicate some of the sites in the city where levees were breached during Katrina. There were a total of 50 sites where failures occurred. The city's naturally low elevation coupled with the failed flood protection system proved a fatal combination. The red areas on the map indicate where deaths occurred (estimated in excess of 1480). A computer-generated simulation of Katrina's flooding in the New Orleans region is available at: www.nola.com/katrina/graphics/flashflood.swf

Hurricane Rita hit just 3 weeks later on the Texas—Louisiana state line, with winds estimated at 120 mph. It caused more coastal erosion, a second evacuation of areas affected by Katrina, localized flooding and additional structural failures. While Rita's impact was most severe in southwestern Louisiana and Texas' upper Gulf coast, New Orleans suffered surges in excess of 8 feet and breaches occurred in some provisionally-repaired levees. These failures caused a second flood of two to three feet in certain neighborhoods, a dreadful replay of Katrina in New Orleans.

Six years later, the extent of the storms' devastation is still shocking. According to the Louisiana Recovery Authority, these storms in combination caused over $200 billion in losses and 1,800 deaths. They destroyed 204,500 homes and 18,750 businesses (Louisiana Recovery Authority, 2006). The storms also eroded over 100 square miles of coastal wetlands, equivalent to what would naturally occur in 50 years. Nineteen of the state's 64 parishes were severely affected by the storms. Some were totally destroyed. New Orleans sustained

Figure 4.   Twin Spans. Photo courtesy of Louisiana Department of Transportation and Development, 2005.

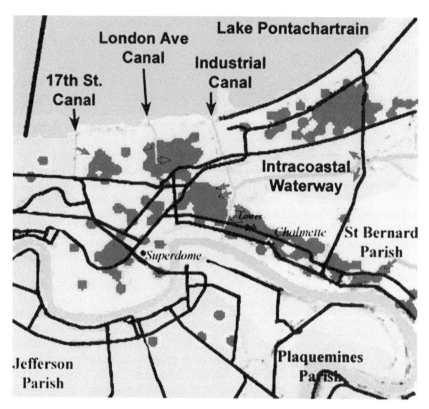

Figure 5. Diagrammatic map of Katrina flood protection failures in the New Orleans Region. Map courtesy of The Times Picayune, 2005.

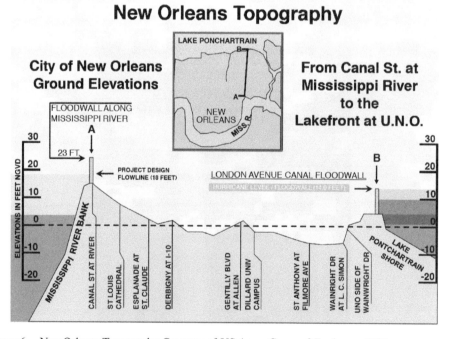

Figure 6. New Orleans Topography. Courtesy of US Army Corps of Engineers, 2005.

Figure 7. Abandoned school in Holy Cross. Courtesy of James Amdal, 2012.

57 percent of the state's total economic loss. Access to the city was severely restricted for five weeks following the hurricanes. Even today, an estimated 60,000 people who lived in the city at the time of Katrina have not yet returned to New Orleans.

4.1 *Grassroots actions and activism became imperative*

Soon after returning to the city, New Orleanians knew what they could do on their own. To put it bluntly, New Orleans was an utter disaster zone. It was eerily quiet. It was almost totally devoid of noise: no birds tweeting, no dogs barking, nothing but a grim silence spread through the city. And Katrina's wreckage was everywhere. Refrigerators stood on the street like giant tombstones. Garbage and downed trees were everywhere. There were unkempt lawns and neutral grounds. Debris littered the streets and clogged sewer drains. Flooded remnants of houses were piled by the curbs. The list went on and on. But into this hell hole many citizens realized that they could provide immediate help to New Orleans in a multitude of ways. The first step was really quite simple: roll up your sleeves, get a tool and go to work cleaning up after Katrina. This grassroots activism was—and continues to be—an important part of the city's recovery. The diverse roles played by volunteers as well as residents in the city's recovery and resurrection was of immeasurable importance and can't be over emphasized. The Katrina Krewe, a knockoff of Mardi Gras terminology, banded together as volunteers to clean the streets, parks, and medians throughout the city. Civic volunteerism by locals, faith-based visitors, non-governmental organizations (NGOs), as well as foreigners, became a necessary response to the immense scale of the damage and the slow pace of the government's response at all levels. Citizens pitched in to enhance or accelerate efforts being made by the city, private contractors, the National Guard and others.

NGOs (i.e. Habitat for Humanity; Global Green and countless others) came to New Orleans to spearhead individual projects. The Musicians Village, in the Upper 9th Ward, sponsored by Habitat for Humanity, is just one example of their lasting contribution. This project

Figure 8.   Debris from Hurricane Katrina. Photo courtesy of James Amdal, 2005.

Figure 9.   Music Village. Photo courtesy of James Amdal, 2012.

is now permanent housing for musicians and, their families who are required to invest "sweat equity" into their residence. In late 2011 the Musicians Village opened the Ellis Marsalis Center for Music as a multi-purpose arts facility for the community and the surrounding neighborhood.

4.2 *The State's response*

In light of the devastation to New Orleans and to southern Louisiana as a whole, Gov. Kathleen Blanco created the Louisiana Recovery Authority (LRA) in mid-October, 2005. It was co-chaired by Dr. Norman Francis, President of Xavier University and Walter Isaacson, former Chairman and CEO of CNN and a native New Orleanian. The LRA was the designated administrator of all federal recovery funds allocated to the state by the U.S. Congress ($10.4 billion). As such, the LRA became key to the state's and city's recovery. The distribution of federal money was contingent upon individual recovery plans submitted by all 19 affected parishes to the LRA for their ultimate approval. In both Louisiana and in New Orleans, creating a comprehensive and inclusive recovery plan became a long and difficult process. This, in large part, was due to a strong sense of "rootedness" that residents felt for their neighborhoods and communities. This sense of being "rooted" to a place often clashed with recovery planning processes that attempted to limit the areas that would be repopulated. This became quite clear during the first LRA sponsored Louisiana Recovery and Rebuilding Conference held in New Orleans November 10–12, 2005 and attended by 650 invitees representing citizens, community leaders, planners, engineers and educators. At one point during the proceedings a facilitator questioned the attendees, "How many of you are from a family that has called Louisiana home for at least three generations?" Over 75 percent indicated yes, an amazingly high percentage. This unique sense of being part of a long line of Louisiana and/or New Orleans families rooted to a specific place became extremely important in the later months and years when recovery strategies were being formulated, vetted and implemented.

4.3 *The city of New Orleans responds*

Preceding the creation of the LRA, New Orleans Mayor C. Ray Nagin created the Bring New Orleans Back Commission (BNOB), a 17-member committee of city leaders, to respond to the devastation and deliver a city-wide recovery plan within 90 days. Its members included 2 college presidents, the Archbishop of New Orleans, attorneys, religious and civic leaders as well as other highly respected members of the business community. BNOB represented a top-down decision-making process but one that encouraged significant citizen involvement.

In order to maximize citizen input, at a time when few residents were actually living in the City, BNOB was organized into committees and subcommittees. These included: land use; infrastructure (flood protection, public transit, criminal justice); culture; education; health and human services; economic development; and government effectiveness. These groups held hundreds of individual meetings over a three-month period and made substantial contributions to the city's initial recovery efforts. Since everyone was focused on the failure of the existing system of floodwalls, floodgates and levees, citizens demanded meaningful improvement to the entire flood protection system. Numerous recommendations made by the infrastructure committee enhanced the overall resiliency of the City and the region. Among these recommendations were:

1. Move the pump stations from their existing interior locations to the lakefront.
2. Install floodgates at the relocated pump stations to protect the existing drainage canals from storm surge.
3. Close the Mississippi River Gulf Outlet, a manmade shipping channel that served as a "hurricane highway" for Katrina's storm surge which devastated St. Bernard Parish, New Orleans East and the Lower Ninth Ward.
4. Reconstruct the Lake Pontchartrain Sea Wall and Lakeshore Drive to reinforce this first layer of New Orleans' flood protection system at the lake.

5. Mandate that the US Army Corps of Engineers complete the New Orleans Primary Flood Protection System to withstand a Category 3 storm.
6. Reform and depoliticize existing parish levee districts by creating 2 reconstituted regional levee districts: Southeast Louisiana Flood Protection Authority – East and Southeast Louisiana Flood Protection Authority – West, with the Mississippi River serving as the natural dividing boundary.
7. Fund and implement significant coastal restoration efforts at the regional and state level.

Historic preservation was another area of intense interest after Katrina. Given that significant portions of the city's historic fabric had been spared massive devastation by the storms, many residents regarded the city's wealth of historic structures, neighborhoods and designated districts as a unique resource for disaster recovery and economic development. They successfully lobbied for expanding the number of designated historic districts; urged leaders to use historic preservation as a foundation for recovery and neighborhood revitalization; suggested that blighted properties be rehabilitated rather than razed; and that existing national preservation programs, such as the Urban Main Streets program, be aggressively adopted in appropriate neighborhoods. The Preservation Resource Center (PRC), a New Orleans based not-for-profit, partnered with The National Trust to open a field office in the city and recruited professional preservationists to assist local advocates in recovery-related tasks, including damage assessments of properties, homeowner assistance in design and construction, and serving on special task forces. The PRC also sponsors annual events (Christmas in October) and programs (Live in a Landmark) to raise citizens' awareness of the rewards of historic preservation on a house-by-house and neighborhood-by-neighborhood basis.

4.4 *Meetings, meetings and more meetings*

As weeks turned into months and eventually years, citizens in New Orleans and in South Louisiana were subjected to countless meetings of all types and sizes with one consistent goal: disaster recovery. Residents felt obligated to attend many meetings. Hundreds, if not thousands, of meetings were held citywide, in individual neighborhoods or disaster-impacted communities, regarding recovery and post-disaster rebuilding. In some neighborhoods these meetings continue today. Despite the overwhelmingly intensive nature of the disaster recovery planning process undertaken in New Orleans, most participants felt these meetings were

Figure 10.   Community Meeting at City Park Pavillion. Photo courtesy of James Amdal, 2006.

well worth the effort. In large part they felt that the ultimate recovery of their neighborhood and the city in general could only be assured by their participation.

4.5   *Outside experts weigh in on disaster recovery*

In the late fall of 2005, a team of volunteer professionals from the Urban Land Institute (ULI), a Washington D.C.-based not-for-profit, was invited to New Orleans to brainstorm with community, business, and academic leaders on the future of New Orleans. This team was composed of planners, landscape architects, mayors, developers, finance experts and public administrators. After a week of analyzing the state of post-Katrina New Orleans, the ULI team presented a recovery plan based on their cumulative experience and expertise. Their recommendations included: shrinking the city's footprint; strategically planning for a reduced population; converting heavily-damaged neighborhoods into open space; and creating a massive redevelopment entity. Most disturbing for many residents, ULI suggested that neighborhoods needed to prove their viability in order to participate in the city's recovery efforts. Finally, they suggested a moratorium be placed on the issuance of any construction permits.

All of ULI's ideas encompassed various aspects of "resiliency". They attempted to prevent or slow the repopulation of areas of the city that were most at risk from flooding. Although reasonable and professionally-sound, ULI's concepts created confusion and fear among many residents, Most of ULI's recommendations were ultimately rejected. ULI's plan paid little regard to the rootedness of many residents in their neighborhoods and seemed to pit one section of the city against another (Mid-City versus New Orleans East, for example). A looming question for all citizens was: "What characteristics made a neighborhood viable?"

On Nov. 19, 2005 the headline in the local daily newspaper announced "4 months to decide." This served as public notice that all neighborhoods seeking recovery funds must prove that they were coming back "safer, stronger and smarter" in post-Katrina New Orleans. Exactly how they were to do this remained to be seen.

Due to political realities, most of the ULI recommendations were rejected almost immediately by the Mayor. He chose instead to adopt a market-driven approach to recovery. With this decision, the idea of imposing terms and conditions on redevelopment and repopulation was dropped from civic discourse. This in turn limited serious discussion about reducing the city's footprint; replacing flood-prone developed areas with water catchment basins/parklands; and clustering housing for greater density in non-flood prone sections of the city.

5   THE LAMBERT PLAN: PLANNING FOR FLOODED NEIGHBORHOODS

In light of the public outcry that ensued following the release of the final BNOB Report in January 2006, which called for the abandonment or delayed repopulation of certain flood-prone sections of the city, Recovery Steering Committees were formed in most city neighborhoods, in large part to prove to the city leaders that their neighborhoods were viable, important and returning. Neighborhoods were grouped into planning districts based on pre-Katrina designations made by the City Planning Commission. Despite its wholesale rejection by the public and most city politicians, the BNOB plan did generate a renewed sense of worth and purpose for neighborhood organizations citywide. But these organizations recognized their need for professional help in developing "credible" recovery plans. Initially, no one knew how that help was going to be provided. During deliberations with the Federal Emergency Management Agency in early 2006, it was determined that FEMA funding could not be used for recovery planning, leaving open the question of who would provide the necessary financial and professional resources to develop neighborhood recovery plans.

Paul Lambert, a Miami-based housing consultant under contract to the New Orleans City Council, came up with the answer. Given his experience with recovery efforts in Dade County Florida after Hurricane Andrew hit in 1992, he knew that unspent Community Development Block Grant (CDBG) funds could be used for recovery planning in the

New Orleans neighborhoods flooded during Katrina. "Dry" neighborhoods (unflooded) were ineligible. Lambert formed a team of local and national professionals that began work in early 2006 on CDBG-funded plans for the city's 40-odd "wet" neighborhoods, outlined in black on the map below. A few neighborhoods elected to conduct their own recovery plans using resources provided by academic institutions, NGOs such as Acorn, and other interested parties.

District 5, depicted by the light olive color to the upper left of the map, constitutes a series of seven neighborhoods immediately city-side of the infamous 17th Street drainage canal. When its city-side floodwall ruptured, catastrophic flooding ensued. Other neighborhoods located immediately south of Lake Pontchartrain were flooded when Katrina's storm surge overtopped the protective levees at the lakefront. District 5 flooded with over 10 feet of water and many homes were filled with 6 feet of mud and debris. Most structures in District 5 were constructed as slab-on-grade after World War II (WWII) and were decimated by the flood-waters. However, in the older neighborhoods (2 are designated historic districts), residences built at the turn of the century suffered less damage. They were constructed on naturally high ground and were built as raised structures (1st floor several feet above grade). They remained relatively "dry" and certainly were more resilient.

Figure 11.   Map courtesy of The City of New Orleans Neighborhood Rebuilding Plan Report. November, 2006.

Today, these neighborhoods are roughly 70 percent to 90 percent repopulated with newly-built and rehabbed houses. Harrison Avenue, the district's commercial corridor, is thriving. A new Hynes Elementary School has opened on its original site. The churches and their school affiliates are flourishing. But the recovery of this district is the direct result of years of dedicated work by citizen activists, strong neighborhood organizations, and significant leadership provided by numerous religious institutions and their congregations.

The importance of religious communities of all faiths cannot be over emphasized when discussing disaster recovery. Immediately after the storm, churches were used as networking centers. They were used as rallying points for parishioners when most means of communications were down. They became beacons of hope for their members and general cheerleaders for neighborhood renewal. They never considered leaving their spiritual home for "greener pastures" outside New Orleans. They committed themselves and their congregations to the long fight for renewal from the very first weeks after the neighborhoods were dewatered and they have never stopped in their efforts.

The District 5 Recovery Steering Committee was created by area neighborhood organizations at the same time the Mayor created the BNOB and the Governor created the LRA. The District 5 Recovery Steering Committee created over 72 different categories for citizen involvement. Meetings were held at least once every two weeks for a period of several months, as attendees added important citizen input into the overall planning process. Keys to the recovery of District 5 were the highly-regimented organization of the Recovery Steering Committee, its member organizations' collective history of advocacy, and residents' relative affluence. Their track record made a real difference when it came time for civic mobilization and political action. Their overriding concern in these early months was convincing themselves, as well as city officials, that District 5 would become a viable neighborhood in the "new" New Orleans.

The final 1,200-page Lambert Plan incorporated 48 individual neighborhood-rebuilding plans and was unanimously adopted by the City Council in November 2006. It included recommendations for projects prioritized based on an implementation timeline that identified the level of importance of each project as determined by the residents and their neighborhood leaders. The total cost for implementing the plan's recommendations was estimated at $4.4 billion. However, because the Lambert Plan could not address "dry" neighborhoods, it could not be used as the city's Recovery Plan to gain access to funds from the LRA. Thus, the Unified New Orleans Plan (UNOP) was conceived as a truly representative recovery plan for the entire city and its respective neighborhoods (both "wet" and "dry").

6 THE UNIFIED NEW ORLEANS PLAN

The Unified New Orleans Plan was a privately-financed planning process that attempted to divorce the planning process from politics as usual. The UNOP included all neighborhoods in the city and sought to incorporate all previous plans into its process and product. These included the BNOB, the Lambert Neighborhood Rebuilding Plans, independent recovery plans created by third party advocates, and Louisiana Speaks—the state's recovery plan. Therefore, it could be, and eventually was, used by the City to access recovery monies from the LRA.

6.1 *A winning tool for demonstrating UNOP's validity: Community congresses*

UNOP's Community Congresses, organized by America Speaks, were largely the key to the plan's success. Eight-hour sessions held on a Saturday in different cities on the same day over three weekends attracted the participation of thousands of residents via a hi-tech, interactive communication tool to allow for instantaneous voting for specific programs, projects or initiatives. The inclusiveness made possible by participation tools that allowed participation by those who had returned to the city and those still in dispersed convinced the LRA that UNOP was the city's most representative recovery plan. Thousands of citizens participated

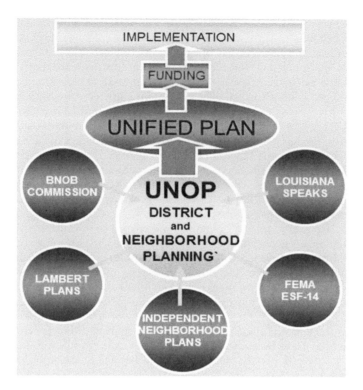

Figure 12.   Organization of Plans Courtesy of UNOP, 2007.

with each other and the Congress facilitators during these meetings. An interactive technique helped convince the LRA that UNOP was the city's recovery plan, based on input from residents and those still in the diaspora. Citizens reached consensus on goals, objectives, policies and projects. This was UNOP's most important contribution: it proved to the LRA that UNOP was the recovery plan for Orleans Parish.

UNOP also analyzed city-wide systems including sewage, water, drainage, power generation and distribution, and communications, as well as all 72 individual neighborhoods with an emphasis on risk-management. Figure 3 illustrates the depth of flooding within the City as established by aerial photography. As you can clearly see, the "Sliver by the River" was spared most of the flooding.

Additionally, UNOP attempted to forecast repopulation using a variety of indicators including utility hookups, mail deliveries, and permit activity by classification. These proved imperfect means of attempting to gauge an uncertain future. In fact a lot remained unknown: the number of residents that would return; where they would choose to live if they were to return; and the city services they would require. Answering these questions with any certainly proved to be very problematic in the end. However, these metrics were useful in approaching an uncertain future.

UNOP, using city-wide assessments of physical, social and civic systems, selected projects and programs to be included in its final list of recommendations. These were broadly prioritized in time (immediate, mid-term, long-term) as they had been in previous recovery and rebuilding plans (BNOB and The Lambert plans). However, it was the community engagement process, not the list of prioritized project, that sold UNOP to the people of New Orleans and to their elected and appointed officials. The costs associated with UNOP's policies and programs were in excess of $14 billion, reflecting its incorporation of all neighborhoods in the city and their collective needs.

A significant part of UNOP required planning for the recovery of the "dry" neighborhoods, where the physical damage was less significant but the economic toll caused by the

Figure 13.   Arial rendering of future VA Hospital Complex. Courtesy of Southeast Louisiana Veterans Healthcare System, 2012.

storm was in some cases dramatic. In approaching two of New Orleans' prime neighborhoods, it became clear that their damage was economic, not physical. The District 1 Recovery Steering Committee incorporated both the Central Business District (CBD) and adjacent French Quarter. These neighborhoods, designated District 1 for planning purposes, recognized a common challenge: the global perception that New Orleans was a neverending disaster with no real future. This meant severely-reduced visitor numbers (tourists and conventioneers) and an extremely weakened economy. However, by working together, their individual interests were joined in mutually-reinforcing plans, programs and projects that benefitted both neighborhoods as well as the city. These included marketing and promotional campaigns by the city and its tourism and convention agency, the state and others with an overall goal to restore tourism. As just one indicator, there are now 300 more restaurants operating in New Orleans than there were before Katrina.

For District 1, UNOP proposed a concentration of performing arts venues at a prime intersection in the CBD. "Broadway South" is now being realized at the upper end of Canal Street, the traditional commercial corridor in the CBD. The 1927 Saenger Theater, an entertainment icon for the City, is being renovated to recapture its historic past while employing cutting-edge technologies for the performing arts. Across the street, the 1947 Joy Theater has been renovated and is now hosting live performances. A new $70 million mixed-use development, primarily residential, has also just been approved by the City Council and groundbreaking will begin in 2012. All of this development has only begun in the last twenty-four months.

UNOP's District 1 plan also called for specific nodes of development with particular characteristics per project. The recent opening of the Bio-Innovation Center, a $60 million incubator for research and development in the medical sciences, heralded the emergence of a new bio-medical research and technology sector. Another key component of this burgeoning economic sector is a new Health Sciences Center that includes a new LSU Teaching Hospital and a new VA Hospital, currently being built at a cost of $2 billion in portions of Mid-City, a neighborhood lakeside of the CBD. These new hospitals have been a source of heated debate for the last several years in the community at-large but they are now under construction.

Although UNOP was expansive in its scale and cost, much of the plan is now being realized. On just one upper CBD corridor, a new $45 million streetcar line is being built, with 100 percent federal funding, with $1.3 billion in new development planned or built along

Figure 14.   Morial New Orleans Convention Center. Photo courtesy James Amdal, 2011.

three blocks of its route (Augustine, 2012). Projects taking shape in its midst include a newly renovated Superdome, the newly-renovated Hyatt Regency Hotel, the Benson Tower and the Saratoga Apartments, a 1950s office building converted into 150 units of downtown rental housing. The CBD's first grocery store opened in mid-October, 2011. Most of these projects have only happened in the last 24 months.

Within District 1 (the Central Business District and the French Quarter) traditional development patterns have been augmented by new centers of investment activity. The Superdome and its immediate environs are seeing over $800 million invested in renovations and new construction within a four-block radius. The Convention Center and the Central Area Riverfront are also seeing additions to their traditional offerings: a new cruise ship terminal at Julia Street, expanded convention facilities at the Morial New Orleans Convention Center and extensive renovations to Riverwalk, a specialty retail center overlooking the Mississippi River. The French Quarter is thriving within its historic context and "toute ensemble."

7   ADDITIONAL RECOVERY INITIATIVES

7.1   *Transportation infrastructure*

To address the damage done to the city's major streets by the floodwaters, the Regional Planning Commission (the Metropolitan Planning Organization for the New Orleans region) in partnership with the Louisiana Department of Transportation and Development and the Federal Highway Administration created the Submerged Roads Program to specifically fund the repair of New Orleans' major arterial roadways. Before the storm, local roads were adequate but after the flood they were horrible and deteriorating at an alarming rate due to the influx of salt water with the floodwaters. This program has been able to repair over 56 miles of

Figure 15.   Bike lane on S. Carrollton, Courtesy of Matthew Rufo, KidsWalk Coalition, 2011.

roadways in the city and significantly added to the local bike path system by specifically desig-
nating bike routes within the newly reconstructed roadways. However, the Submerged Roads
Program could not address the needs of the minor roadways such as neighborhoods streets.
How these repairs will be funded remains another unanswered question post-Katrina.

   Attention has also been given to expanding options for mobility with a special emphasis on
improving and enlarging alternative transportation systems, specifically for pedestrians, the
mobility- impaired and cyclists. Through a coordinated plan that addresses multiple needs
throughout the city, New Orleans has grown its bicycle network from a meager 5 miles in
2004 to over 50 miles in 2012 (Jatres, 2012). From 2010 to 2011, bicycle use increased 20 percent. This network includes bike lanes within roadways, shared-lane markings or "sharrows"
and bike trails. In the coming months New Orleans will begin the construction of the Lafitte
Greenway. This 3.1-mile project is converting an abandoned freight rail right-of-way into a
multi-functional linear green space that will accommodate a variety of users from toddlers to
the elderly. Walkers, cyclists, skateboarders, rollerblade aficionados as well as young families
with strollers will use this new amenity. This project, long a dream for a core constituency,
is now being heralded as the ultimate urban renaissance: turning an abandoned industrial
eyesore into a community asset. When complete it will serve multiple neighborhoods along
its path, from the French Quarter to Bayou St. John in Mid-City.

7.2   *Home elevation*

On another front, after the floods, elevating houses became another option for homeowners
to consider, especially in neighborhoods that were built after WWII in newly developed areas
within the city that are extremely vulnerable to flooding. These houses were typically built
as "slab-on-grade" and present costly challenges for the homeowner if they choose to raise
them off the ground. In many instances house raising has created visually disturbing results,
especially when the raising is taken to extremes.

Figure 16.   Houses Raised in Upper Gentilly, Courtesy of James Amdal 2011.

The original elevation program provided up to $30,000 per residence for "house raising". However, this amount was recently determined to be inadequate and was increased to provide residents with the funds necessary to properly raise their homes. Another problem with this program was its timing.

The original "Road Home" program that provided funds for rehab or reconstruction of damaged homes was awarded several years before the elevation program was activated. This complicated the entire rehab and reconstruction funding decision for homeowners. Equally troubling, abuses are now being reported of unqualified contractors being hired for these projects. Although paid through state administered programs, fraud claims are common and becoming more so. Finally there are no design standards for house raising, which can lead to unintended consequences. Many individual homes look out of place when taken in context of neighboring houses and the overall streetscape.

7.3   Levees and flood protection

Since Katrina, the US Army Corps of Engineers (the Corps) has spent over $14.4 billion upgrading, repairing and completing the New Orleans Area Hurricane and Storm Risk Reduction System. This was and remains the city's most important resiliency component. One major project built in response to Katrina is the surge protection barrier in Lake Borgne, which cost $1.1 billion. When complete in late 2012, the projects undertaken by the Corps will provide protection for a Category 3 storm for the greater New Orleans region. However, many believe this system should offer protection from a Category 5 storm but the costs are astronomical and currently this upgrade is not being seriously pursued. Another unintended consequence of the storms of 2005 has been their positive impact on the local economy. These flood protection projects, as well as the rest of our recovery activities, have shielded the New Orleans region from the economic ills currently afflicting the United States. We became an isolated bubble of reconstruction activity post-Katrina.

Figure 17.   Lake Borgne Protection Barrier. Courtesy of US Army Corps of Engineers, 2011.

Coastal restoration has also emerged as a priority within the state and the region. Spear-headed by the America's Wetlands Foundation, a well recognized NGO, "the passage of the 2012 Coastal Master Plan represents a nationally significant comprehensive systems approach to coastal restoration and protection of one of the nation's most critical natural resources.... The State is doing its part to secure important economic and environmental assets that serve America by providing abundant energy, seafood and agricultural resources, as well as world's largest port and navigation systems that underpin the U.S. economy and economies of 31states served by the Mississippi River." according to King Milling, Chair AWF (Ports Association of Louisiana, 2012).

7.4   *Education*

By 2016, every public school student in New Orleans will be attending a new or renovated school, designed and constructed with resilience as a core requirement. This $2 billion rebuilding program is being administered by the Recovery School District, a state agency that replaced the highly politicized pre-Katrina Orleans School Board. An alternate program of charter schools, largely administered by parents and teachers, has proven extremely successful in post-Katrina New Orleans, to the delight of local and national advocates. Fully 78 percent of today's public school students are being educated in a charter school, making New Orleans a model for national public education reform.

7.5   *Public housing*

Public housing has also been transformed post-Katrina into mixed-income developments by an aggressive partnership between the Department of Housing and Urban Development, the Housing Authority of New Orleans and a number of local and national private sector developers. Before Katrina, more than 5,000 families lived in public housing, but today only one-third of these families have returned to the newly built replacements. Former residents have mixed emotions about their new neighborhoods with some decrying the higher rents and utility bills. Others miss their former neighbors who have not returned. It remains to be

Figure 18. Lake Area New Tech Early College High School in Gentilly. Photo courtesy of James Amdal, 2012.

Figure 19. Faubourg Lafitte. Photo courtesy of James Amdal, 2012.

seen if this new model of public housing will fare better than the model it replaced. However, in all cases, these new developments have incorporated resiliency into their design and construction.

7.6 Entrepreneurialism

Post-Katrina, New Orleans has become a hot spot for young entrepreneurs. Many were first drawn to the city by the disaster, but many have stayed and prospered. During a UNOP District 1 Recovery Steering Committee Meeting, a successful marketing executive and member of the committee noted that the new Creative Class residents in New Orleans "can be anywhere. Their businesses are lap-top based, but they have chosen New Orleans due its limitless opportunities for growth and development." The Idea Village, a local Not-For-Profit founded in 2000, has emerged post-Katrina as a nexus for very successful entrepreneurial initiatives. Recently, The Wall Street Journal referenced this latest New Orleans phenomenon in the article *A Tech Buildup on the Bayou*. According to the article "Companies from start-ups to large-established entities are expanding or setting up shop here, drawn by a state tax credit that is offered to digital-media and software firms... New Orleans stock of tech jobs grew 19% from October 2005 to April 2012, compared with 3% nationally." (McWhirter & Dougherty, 2012).

7.7 Crime, blight and other problems

Despite our post-storm recovery, long-standing problems still persist in New Orleans. Crime remains an on-going problem for residents and visitors alike. According to figures recently released by the Greater New Orleans Community Data Center, currently there are over 48,000

Figure 20. Lake Area New Tech Early College High School in Gentilly. Photo courtesy of James Amdal, 2012.

blighted and/or vacant structures and lots in the city. New Orleans ranks number 2 in the US for income disparity, and the 2010 British Petroleum deep-water oil spill is still affecting the local and state economy.

Meantime, despite ongoing neighborhood investment, certain sections of the city stand as unfortunate reminders of the devastating effects of the storms. The Lower 9the Ward is still struggling after six years of planning, advocacy, and major investments by foundations and support organizations. The Make It Right Foundation is in the process of building 150 environmentally-friendly homes in the neighborhood that was decimated by Katrina's flood-waters. The new houses are elevated 8 feet and feature Energy-Star windows and appliances, formaldehyde-free cabinets and paints free of VOCs. The ultimate success of this initiative is still up in the air in spite of its laudable intent and significant investment. Retail and institutional anchors have been slow to reemerge after the storm. However, it was recently announced that a new 25,000 square foot grocery would be built in the neighborhood to serve area residents.

8 CONCLUSION

In February of 2007, Bob Hebert, a columnist for the New York Times, noted that New Orleans was like the fairy tale character Humpty Dumpty who fell off the wall and nobody knew how to put him back together again. At the time, his assessment about the city was accurate: "A great American cultural center like New Orleans was all but washed away, and no one knows how to put it back together again." (Herbert, 2007). Since then, much progress has been made, but serious problems still persist. The most pressing issues include the city's high crime levels and the vast numbers of blighted or abandoned housing that sit dormant in many neighborhoods.

The post-Katrina recovery of New Orleans presents a unique opportunity for others to learn from our successes and failures. Today, New Orleans may be the best laboratory in the world for academic and applied research in the ever-expanding disciplines of disaster recovery and urban resilience. There exist countless avenues of investigation: from public health to economic revitalization. In spite of overwhelming odds, New Orleans has demonstrated success in neighborhoods throughout the city. Each offers a unique view on disaster recovery from many perspectives: the individual citizen, the neighborhood leader or organizer, and city, state or national administrator or policy makers in the public and/or philanthropic realm.

Resilience has many different faces in post-Katrina New Orleans: the physical, the social, the historic and the organizational. The city has excelled in some areas more than others, but successes have been achieved in each of these categories. New Orleans over the last six years has been re-envisioned, rebuilt and resurrected from the floodwaters of 2005. Post-Katrina, the city clearly has much to offer the international community in understanding and learning from our efforts in disaster recovery and urban resilience.

REFERENCES

Augustine, J. (2012, February 15). *Regional Transit Authority Transportation Investments Generate Economic Growth*. Retrieved May 31, 2012, from The White House: http://www.whitehouse.gov/blog/2012/02/15/regional-transit-authority-transportation-investments-generate-economic-growth
Jatres, Dan. (2012, June). *Regional Planning Commission*. Interview.
Herbert, B. (2007, February 22). From Anna to Britney to Zawahri. *The New York Times*, A23.
Louisiana Recovery Authority. (August 2006). *Hurricane Katrina Anniversary Data for Louisiana*. Louisiana's Media Center. Louisiana: Louisiana Recovery Authority.
McWhirter, C., & Dougherty, C. (2012, June 8). A Tech Buildup on the Bayou. *The Wall Street Journal*, A3.
Ports Association of Louisiana. (2012, June 6). News from the Docks. *13(6)*. New Orleans, Louisiana.

Resilience and Urban Risk Management – Serre, Barroca & Laganier (eds)
© 2013 Taylor & Francis Group, London, ISBN 978-0-415-62147-2

Assessing the resilience of urban technical networks: From theory to application to waste management

H. Beraud, B. Barroca & G. Hubert
Université Paris Est—Marne la Vallée, Urban Engineering Group, Paris, France

ABSTRACT: Waste management system plays a leading role in areas' capacity to re-start after flooding, as their impact on post-crisis management can be very considerable. Therefore, they have to be resilient for the town to be able to survive and live on. The purpose of this article is to reflect how to define and assess the resilience of the waste management network.

1 INTRODUCTION

Resilience improvement strategies have taken a preponderant place in risk management policies. This evolution is in keeping with a paradigm shift process, which has changed from management that is essentially focussed on reducing the impact of a hazard to management that looks more closely at questions on how to get a territory back on its feet again. The objective is no longer just to fight against floods, but also to learn how to live with them. Reasoning in terms of resilience enables this change in focus to take place. Resilience is defined as the persistence of relations within a system and as a measure of the ability to absorb and integrate changes in its component elements (Holling, 1973). It is certain that systems are permanently evolving. As a result, they are not characterized by a state of balance but by the general stability formed by maintaining them in operation. In this way, whenever a disturbance arrives, either the system is capable of integrating it with a number of disturbances that do not put its viability into question, in this case it is resilient; or the system is incapable of doing so, in which case it is deteriorated at more or less long term by the change in its structure (Sanders, 1992). Therefore, resilience suggests that the system takes a proactive position towards risks. In this way, it is the system's internal characteristics that enable it to react in the face of a risk (Pelling, 2003). Therefore, the capacity to react, and over and above the capacity to adapt, appears to be the central point for characterizing a system as being resilient. For our subject, we are retaining three aspects for characterizing a system as being resilient: (1) its capacity to react, (2) the reference to a return to an acceptable operating condition for it to be maintained and, lastly, (3) its capacity to adapt and to learn for the future from catastrophes of the past.

By their organisation and the multiplicity of their functions, towns and cities can be considered as being complex systems. They are territories that are particularly vulnerable in the face of flooding. Damage to them may weaken the way the territory globally operates and put a country's economy into peril. Therefore, improving their resilience would appear to be primordial. A resilient town is a town capable of guaranteeing its continuity and adapting itself to changes in the surrounding environment. To do this, it relies on a self-organisation capacity, that is to say a capacity to permanently adjust its behaviour depending on interactions both inside it and with the external environment (Pumain et al., 1989). Urban technical networks occupy a basic position in this capacity (Campanella, 2006, Sanders, 1992, Pelling, 2003). They play the essential role as a support for spreading information, materials, and decisions that enable the territory to operate and make its resilience possible. Therefore, they have to be resilient for the town to be able to survive and live on.

The present communication aims at reflecting on how to define and assess the resilience of urban technical networks via the study of a central network for providing towns with resilience against flooding, a network that has been only little studied, the waste management network. To do this, the first thing to be done is to explain and analyse the notion of resilience in urban technical networks. Secondly, an analysis will be made of specific resilience elements in waste management networks. Lastly, it will be shown how resilience can be assessed by analysing the waste management networks of la Faute-sur-Mer and l'Aiguillon-sur-Mer perturbed by the passage of hurricane Xynthia.

2 THE RESILIENCE OF URBAN TECHNICAL NETWORKS OR THE IMPORTANCE OF THE NETWORK'S STRUCTURE

An "Urban technical network" is the name given to a service that has been made available (waste collection, drinking water supplies, energy supplies, public transport, etc.) based on an organization that operates on a territorial level and on an infrastructure. By considering them as complex systems, they can be defined as being the link between a service network corresponding to the network's mission and the reasons for its existence and its organisation (functional aspects) and a support network that represents the infrastructures, the equipment, the resources required for providing the service (Blancher, 1998).

Networks are structured by a set of so-called "occasional" elements, related to each other by linear elements, flow supports and forming, in this way, a meshwork. Each occasional element corresponds to a distinct, different entity. They are support points for the network's activity (bus garages, electricity production sites, sewage works, waste production sites, etc.) but also sites where the persons involved (network administrators, authorities, etc.) think out their actions, that is to say, the place where individual or collective will for relations with another point or person involved is born (Dupuy, 1991). The meshwork corresponds to the infrastructure (roads, railways, piping, etc.), which are the supports for the flows required to provide the service. This meshwork creates a territorial zone that corresponds to the grid or the structure of the network (Gleyze and Reghezza, 2007).

The arrival of floods on a territory may cause a dysfunction, even a break, in an urban technical network due to the difficulties, even the impossibility, for it to carry out its normal missions. This risk of the network's functional dysfunction is related to damage to the support network (damage to equipment, linear infrastructures, etc.). For example, if sewage works are flooded, they cannot carry out their normal mission for providing uncontaminated water. However, modifications to the network's structure generally accompany material deterioration to infrastructures such as piping. As flows can no longer take their normal route, they may be led to using a different one. The arrival of floods, and therefore the unavailability of certain nodes in the network, may make the whole structure of the network inefficient, deteriorating its conditions of use at the same time (Gleyze, 2005). Therefore, the structure's architecture may or may not contribute to reducing or increasing the network's vulnerability. In this perspective, and depending on the damage made to certain elements, it determines the deterioration of the service made available (Lhomme et al., 2010). Therefore, the resilience of an urban technical network must take account of these three levels of networks.

Firstly, resilience is understood as being the capacity of the service network to survive in an operating condition acceptable to society. To do this, all the component elements in the support network must be able to react and adapt themselves to any modifications and dysfunctions produced by the hazard. However, it is the structure itself, comprised of the different infrastructures, which will make the network efficient or not, and therefore capable of reacting and adapting itself to disturbances and being able to "return to normal". Therefore, the network structure's permanence depends on its capacity, via its infrastructure, to "fill in empty gaps" (the possibility of creating loops, redundancy of networks), but also to adapt and regulate itself (Dupuy, 1985). For example, a network with a strong relationship[1] and strong connectivity[2] will favour the

1. Relationships materialize the fact that points may or may not be connected to networks. A strong relationship means relations between numerous points on the network (Dupuy, G. 1985. *Systèmes, réseaux et territoires: principes de réseautique territoriale,* Paris, Presses de l'ENPC).
2. Connectivity assesses the possibilities of direct and alternative relations between several points on a network (Ibid.).

system's capacity to adapt and react, and therefore its resilience. In this type of network, if the main flow route becomes unavailable, with a high density of meshwork, the flow can always take another route. Therefore, reflecting on the resilience of a network means questioning whether the relations that enable the mission to be carried out can be maintained, and, thereby, whether the flow nodes and supports will be able to react and adapt themselves to any change caused by a hazard. Therefore, to do this, an analysis must be made of the capacity of the support network and its structure to recover and adapt themselves. This initially requires for the fragility of the support network to be analysed, i.e. its potential damage level, and then for the redundancy and looping capacities of the network structure to be measured depending on the damage level.

3 THE SUPPORT NETWORK AT THE HEART OF THE WASTE MANAGEMENT NETWORK'S RESILIENCE

Even though it has rarely been studied, the resilience of a waste management network proves to be essential for a territory to operate correctly. It is certain that disturbances in its operation, even total stoppages, can have extremely detrimental consequences in terms of health and public sanitation and for putting the territory back into action. Now, as we shall see, the impact of flooding on this network is by no means negligible.

The waste management network is organized in the form of waste management procedures[3]. Different entities involved (producers, administrators, players in charge of control, etc.), regulations and specific infrastructures correspond to each procedure. Globally, like the other networks, a waste management network is characterized by a use (service network), an infrastructure (support network) and a territorial area. The service network corresponds to the organization set up for enabling it to carry out its primary mission—to manage waste in a way suited to the type of flow, for sanitation and public health purposes. It is organized into "Waste management procedure" systems, which are broken down into five sub-systems that correspond to the five stages to be found in waste management: Collecting, storing, transforming, burying and upgrading. This "procedure" system is related to elements from the outside environment through which and for which it exists: producers of waste, society in general, the world market, regulatory authorities (State, EU, services, etc.) companies that are users of secondary raw materials and networks (water, sewage, electricity, gas, thoroughfares, telecommunications, etc.) (Beraud et al., 2011b). This organization is based on an infrastructure that is specific to every sub-system, comprising a road system, sites for producing, collecting, treating, eliminating and upgrading waste and sites for organizing the activity (operational centres and decision-making points). Each one of these sites can be considered as being a node in the waste management network. Each node possesses its own organization and its own resources for enabling it to carry out its mission. Relations borne by flows are established between nodes. It is this set-up, called the support network, which enables service network objectives to be attained. Creating relations between nodes by means of flows creates a framework that is a part of the territory, but which, contrary to other networks, is not permanently fixed. Waste flows are not in fact "territorialized". Even if the nodes from and to which they circulate are fixed, the routes they take may change.

The consequences of flooding on a waste management network are not negligible: dysfunctions in collection, treatment or organisational infrastructures (due to personnel, sites or even equipment being unavailable), but also changes in the node's internal structure (finding new partners, new sites, etc.) even changes in its main function (evolutions in the priority missions of the waste management network, …) due to evolutions in the nature and quantity of waste flows (Beraud et al., 2011a). Disturbances to waste management networks are primarily linked to damage or non-adaptation of infrastructures[4]. The importance of the

---

3. The "procedure" represents all the activities related management of a specific type of waste. For example, fermentable waste is collected, stored and treated following a process and regulations, which are specific to the nature and characteristics of this type of waste. Management of this type of waste forms a waste management procedure.
4. Waste treatment sites, operating centres, decision centres for the activity or equipment damaged by flooding or incapable of facing up to the new waste production.

Table 1. Comparison of resilience criteria between urban technical networks and waste management networks.

Network element	Resilience of an urban technical network	Resilience of a waste management network
Service network	Return to acceptable operating conditions, which enable it to survive at long term	Return to acceptable operating conditions, which enable it to survive at long term The capacity to mobilize resources that will enable post-flood waste to be handled.
Support network	The infrastructure's capacity to react and adapt itself	The capacity to react and adapt itself to variations in the infrastructure's operation and quantitative and qualitative variations in waste flows.
Structure	The capacity to adapt and react (looping, redundancy)	The structure's capacity to adapt and react.

network's structure is less significant in this case, as it is not permanently fixed. What in fact matters above all else is that waste is transported from its point of production to its point of treatment irrespective of the route. If roads are impracticable, as they are flooded or blocked by debris, a replacement route can take over the relay, just as long as there are other routes available. Therefore, in this case, the structure of the network plays a secondary role.

Therefore, a resilient waste management network must be capable of remaining in operation at acceptable levels. The term "acceptable" means that the missions for which the network was created and, consequently, the functions that enable it to carry out these missions[5] must be maintained (Beraud et al., 2011b). For this reason, the network must be capable of reacting and adapting itself not only to variations in structures' availability and operational capacity, but also, and above all, to variations in the production of waste (quantitative and qualitative). This last characteristic is one of the main specificities of waste management networks when compared with other urban technical networks. This is because this network must get back into action and work much more intensely than under normal conditions, as it has to handle waste in considerably larger quantities and in new forms. The specific aspects of resilience in waste management networks are summarized in the following table (Table 1).

Therefore, assessing the resilience of a waste management network really means questioning the capacity of infrastructures to adapt and react. For example, are treatment, storage and operating sites in zones liable to flooding? If so, to what extent is the administrator capable of mobilizing other treatment or storage resources? Are management resources sufficiently dimensioned to face up to any evolution in waste production? By answering these questions, the resilience of the waste management network can be measured, as we shall see below.

4 RESILIENCE OF THE WASTE MANAGEMENT NETWORK AT LA FAUTE-SUR-MER AND L'AIGUILLON-SUR-MER FOLLOWING THE PASSAGE OF HURRICANE XYNTHIA IN 2010

Following the flooding that resulted from the passage of hurricane Xynthia in February 2010, The equivalent of twelve years' production of waste[6] was produced in one single night in the

---

5. By this, we refer to the missions carried out by the waste management network defined in the context of a functional analysis. This method enables us to understand the way in which a system operates by characterizing its structure, its environment, its functions, together with the conditions under which it operates (Villemeur, A. 1988. *Sûreté de fonctionnement des systèmes industriels. Fiabilité—Facteurs humains—Informatisation,* Paris, Eyrolles, Coll. Etudes et recherches d'Electricité de France.).
6. Or about 7 000 ton of waste in the two municipalities (Robin Des Bois 2010. Les déchets de la tempête Xynthia.).

municipalities of La Faute-sur-Mer and L'Aiguillon-sur-Mer. The municipalities, in charge of collecting household waste, and the area household waste syndicate, Trivalis, the main participants in the waste management network for the territory concerned had to face up to this new production, whilst trying to cope with "normal" waste at the same time. Despite certain dysfunctions (pollution of temporary storage areas, remnants of high-water levels in a natural environment, difficulties in sorting certain highly dispersed waste, etc.), the waste management network appears to have reacted relatively well in the face of the catastrophe. Even so, can it be qualified as being resilient?

As defined above, a resilient waste management network is capable of remaining in an acceptable operating condition. To do this, its support network must be able to react and adapt itself both to variations in the availability and operating capacity of infrastructures and also to waste production. What is the situation for the network presented here?

In view of the resources mobilized to return to normal after the catastrophe (Table 2), globally, it would appear to have been capable of reacting and adapting to these variations.

This capacity can be explained by the conjunction of several different factors. On the one hand, whereas the whole zone of action of the regional waste treatment syndicate,

Table 2. Answers to flooding provided by the local waste management network.

Network elements	Resilience criteria for waste management networks	The answer to flooding provided by the local waste management network
Service network	Return to acceptable operating conditions, which enable it to survive at long term The capacity to mobilize resources that will enable post-flood waste to be managed.	Mobilizing the whole territory of action Reorganizing collecting and treatment procedures
Support network	The capacity to react and adapt to variations in the infrastructure's operation and quantitative and qualitative variations in waste flows	Redistributing material resources available within the "normal" organization of the waste management network. Mobilizing external resources notably in terms of waste collection and temporary storage sites.
Structure	The structure's capacity to adapt and react.	Nothing to report

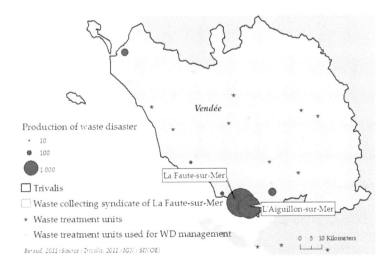

Figure 1.   The consequences of hurricane Xynthia on the "Vendée" waste management network.

Trivalis, is the whole "Vendée département", only the municipalities of La Faute-sur-Mer and L'Aiguillon-sur-Mer suffered from the hurricane severely (Figure 1).

This exceptional situation enabled them to benefit from considerable resources provided by Trivalis, but also from the solidarity of local entities involved (entrepreneurs, farmers, etc.) and from neighbouring municipalities. It is highly possible that if the number of municipalities stricken by the hurricane had been greater, these resources would have been more dispersed and therefore less efficient. Moreover, few waste management infrastructures were damaged (notably treatment units).

Nevertheless, even if the waste management network was capable of recovering from the flooding caused by hurricane Xynthia, it is not certain that this would have been the case if flooding had occurred over a larger area on the territory concerned. It is certain that one of the capacities available to the network for managing the event was that dysfunctions only concerned a small part of its complete area of action. Its capacity to react and adapt was made easier by this situation. In this territorial context, the waste management network was relatively resilient. However, a resilient system must also be capable of learning from past catastrophes. It must ceaselessly improve itself to maintain itself at long term. Therefore, it is necessary for the waste management network on the territory concerned to learn from the way the catastrophe was managed.

5 CONCLUSION

Characterizing the resilience of urban technical networks to flooding proves to be essential for a better understanding of urban resilience. For from being as clear as electricity or drinking water supplies, waste management is a network that should not be neglected to enable territories to get back on their feet again as quickly as possible. Therefore, working on its resilience appears to be indispensable. The network's resilience criteria are somewhat different from those brought to light for urban technical networks in general. As a result, on the one hand, the support network's capacity to adapt and react is essential. Indeed, this network has to face up to any dysfunction in the infrastructure (caused by damage), but also, and above all, to new waste being produced. Moreover, the part played by the structure is less significant as flows are not limited to a fixed meshwork. They are relatively mobile and can be adapted to traffic conditions. For this reason, assessing the resilience of a waste management network comes down to questioning infrastructures' capacity to react and adapt with regard to the damage they have undergone and evolutions in the quantities and the quality of the waste produced. Applying these resilience criteria to the municipalities of La Faute-sur-Mer and L'Aiguillon-sur-Mer has shown that this capacity is mainly due to the difference in scale between the network's total area of action and the area that suffered from the catastrophe. In fact, the two municipalities that were flooded after the passage of hurricane Xynthia were able to face up to the catastrophe because the zone covered by the waste management network was much larger than the zone struck by the catastrophe.

REFERENCES

Beraud, H., Barroca, B. & Hubert, G. 2011a. De la nécessaire prise en compte du réseau de gestion des déchets dans les stratégies d'amélioration de la résilience des territoires urbains aux inondations. *Sociétés et catastrophes naturelles.* Orléans.

Beraud, H., Barroca, B., Serre, D. & Hubert, G. 2011b. Making urban territories more resilient to flooding by improving the resilience of their waste management network. A methodology for analysing dysfunctions in waste management networks during and after flooding. *ICVRAM 2011 and the International Symposium on Uncertainty Modeling and Analysis, ISUMA 2011.* Hyattsville.

Blancher, P. 1998. *Risques et réseaux techniques urbains,* Lyon, Certu, Coll. Débats: Environnement.

Campanella, T. J. 2006. Urban resilience and the recovery of New Orleans. *Journal of the American planning association,* vol. 62, n° 2, 141–146.

Dupuy, G. 1985. *Systèmes, réseaux et territoires: principes de réseautique territoriale,* Paris, Presses de l'ENPC.

Dupuy, G. 1991. *L'urbanisme des réseaux: théorie et méthodes,* Paris, Armand Colin, Coll. U. Géographie.

Gleyze, J.-F. 2005. *La vulnérabilité structurelle des réseaux de transport dans un contexte de risques.* Doctorat sous la direction de, Université Paris 7 – Denis Diderot.

Gleyze, J.-F. & Reghezza, M. 2007. La vulnérabilité structurelle comme outil de compréhension des mécanismes d'endommagement. *Géocarrefour,* vol. 82, n° 1–2, 17–26.

Holling, C. S. 1973. Resilience and stability of ecological systems. *Annual review of ecology and systematics,* Vol. 4, 1–23.

Lhomme, S., Serre, D., Diab, Y. & Laganier, R. 2010. Les réseaux techniques face aux inondations ou comment définir des indicateurs de performance de ces réseaux pour évaluer la résilience urbaine. *Bulletin de l'Association de géographes français. Géographies,* n°4, 487–502.

Pelling, M. 2003. *The vulnerability of cities. Natural disasters and social resilience,* London, Earthscan.

Pumain, D., Sanders, L. & Saint-Julien, T. 1989. *Villes et auto-organisation,* Paris, Economica.

Robin Des Bois 2010. Les déchets de la tempête Xynthia.

Sanders, L. 1992. *Système de villes et synergétique,* Paris, Economica, Coll. Villes.

Villemeur, A. 1988. *Sûreté de fonctionnement des systèmes industriels. Fiabilité—Facteurs humains— Informatisation,* Paris, Eyrolles, Coll. Etudes et recherches d'Electricité de France.

Resilience and Urban Risk Management – Serre, Barroca & Laganier (eds)
© *2013 Taylor & Francis Group, London, ISBN 978-0-415-62147-2*

Urban technical networks resilience assessment

S. Lhomme
Université Paris-Est (EIVP), Paris, France
Université Paris Diderot, Paris, France

D. Serre
Université Paris-Est (EIVP), Paris, France

Y. Diab
Université Paris-Est (LEESU), Paris, France
Université Paris-Est (EIVP), Paris, France

R. Laganier
Université Paris Diderot, Paris, France

ABSTRACT: In France, as in the rest of Europe, river floods have been increasing in frequency and severity. In the same time, the total urban population is expected to double from two to four billion over the next 30 to 35 years. These circumstances oblige to manage flood risk by integrating new concepts like urban resilience. The main aim of this paper is to illustrate that networks resilience represent an important issue for urban resilience analysis. Then, two main methods designed for assessing networks resilience are presented: methodology for modeling networks interdependencies and indicators for assessing redundancy. Finally, a Web GIS, implementing these two main methods, is presented.

Keywords: urban resilience, technical networks, Web GIS, interdependency modeling, redundancy indicators

1 INTRODUCTION

The total urban population is expected to double from two to four billion over the next 30 to 35 years. Growing rate is equivalent to the creation of a new city of one million inhabitants every week, and this during the next four decades. In the same time, in France, as in the rest of Europe, river floods have been increasing in frequency and severity. Flood is one of the major natural hazards that has caused loss of lives, significant economic damage, pollution on the nature and the built environment, loss of cultural heritages, and even caused community disorder and health problems (Ashley et al. 2007). In fact, quick urban development, coupled with technical failures and climate change seems to have increased flood risk and corresponding challenges to urban flood risk management (Ashley et al. 2007, Nie et al. 2009). These circumstances oblige to manage flood risk by integrating a new concept: the urban resilience. This concept has emerged because a more resilient system can be considered to be less vulnerable to risk and, therefore, more sustainable (Klein et al., 2003).

The concept of resilience is used in many disciplines (like psychology, economy, geography...), but for risk management this concept is relatively new, especially concerning natural hazards. It is a multidisciplinary concept. Henceforward, resilience is "fashionable", as much with scientists as with the administrators and international authorities in charge of preventing disasters. Comfort et al. (2010) call it a "buzzword" and see its consecration with September 11 and Katrina. The abundant use of the concept, especially in social sciences, does not always come with a solid theoretical base. The word then becomes a holdall word used with a variety of meanings, just like other fashionable concepts (durability, governance…) that are often used in relation to it (Aschan 1998, Gallopin 2006). Resilience became polysemic and Klein et al. (Klein et al. 2003) consider resilience as an "umbrella concept".

Otherwise, every day millions of people in Europe and elsewhere benefit from the development of a highly sophisticated network of essential infrastructure systems to sustain their communities (Demsar et al. 2008). Responding to the pace and economics of modern life, the system is expected to work without failure and the level of tolerance towards traffic jams, delays in the railway system and power supply breakdowns becomes lower (Lhomme et al. 2011). This essential infrastructure is considered as critical infrastructures. Natural disasters, such as a flooding, can cause the fail of entire lifelines in a city or larger area. Even if the flooding is of minor scale it can cause severe damage to infrastructure and critical buildings in which the network control units are located. Nowadays, multiple networks that innervate cities are particularly sensitive to flooding, through their structures and geographic constraints (Murray & Grubesic 2007). Because societal functions are highly dependent on networked systems and the operability of these systems can be vulnerable to disasters, there is a need to understand how networked systems are resilient.

The main aim of this paper is to illustrate that networks resilience represent an important issue for urban resilience analysis. Then, the objective is to determine methods and tools for networks resilience analysis. That is why urban resilience and the importance of networks for urban resilience will be introduced in section 2. Then, section 3 will focus on two main methods allowing urban networks resilience analysis. To conclude, a prototype designed in order to analyze networks resilience will be presented in section 4.

2 NETWORKS RESILIENCE AND URBAN RESILIENCE

In this research, number of disciplines using resilience concept has been studied in order to well understand and to define exactly this concept concerning urban flood management. Indeed, the abundance of definitions of disaster resilience and the fact that this concept is shared by many disciplines makes it difficult to have a common definition. It appears that resilience is usually used in the continuity of existing terms in these various disciplines (Fig. 1). Polysemy implies a rigorous use of the words, and in order to use resilience in an operative way, it is important to keep in mind the theoretical problems and the methodological limits that come with the different meanings, while paying attention to the inherent contradictions of certain uses.

2.1 Resilience definition

Derived from ecology, the concept of resilience was firstly defined as "the measure of the persistence of systems and of their ability to absorb change and disturbance and still maintain

Figure 1.   Resilience and related concepts in different disciplines (Lhomme et al. 2010).

the same relationships between populations or state variables" (Holling 1973). The concept of resilience has a rich history (Folke 2006), sometimes with a considerable stretch from its original meaning (Gallopin 2006). Thus, Folke identifies a sequence of resilience concepts in ecology, from narrow to broad: engineering resilience, ecosystem resilience, social-ecological resilience.

The most important development over the past thirty years is the increasing recognition across the disciplines that human and ecological systems are interlinked and that their resilience relates to the functioning and interaction of the systems rather than to the stability of their components or the ability to maintain or to return to an equilibrium state (Klein et al. 2003). This social-ecological resilience refers to the concept of "panarchy" or cross-scale dynamics and interplay between nested adaptive cycles (Gallopin 2006). Currently, Resilience Alliance, a research organization, proposes very interesting ways of researches. Nowadays, resilience can be considered as *"the magnitude that can be absorbed before the system changes its structure by changing the variables and processes that control behavior"* (Gunderson & Holling 2002).

Urban resilience is more often defined as *"the capacity of a city to face devastating event reducing damage at minimum"* (Campanella 2006). This definition emphasizes the operational aspect of resilience that would tend to reduce the damage caused by a disturbance. But, in this context, resilience doesn't appear really as a new concept, because city managers tend to reduce damages since several decades. Moreover, this definition of resilience can't be linked directly with others disciplines which use resilience concept. So, thanks to the study of many definitions in different disciplines, urban resilience has been defined in this research as *"the ability of a city to absorb disturbance and recover its functions after a disturbance"*. In other word, resilience is *"the ability of a city to operate in degraded mode and to recover its functions whereas some urban components are disrupted"* (Lhomme et al. 2011).

The polysemy of the concept resilience is not a problem *per se*: it is even productive in terms of heuristic and methodological issues (Folke 2006). So, the objective here is not to impose a definition of urban resilience. Resilience definition must be considered in this research as a starting point—based on a literature review in order to be rigorous—which opens new perspectives to get over situations by giving hope in the existence of other solutions that need to be searched for. In fact, throughout semantic debate, the most important concerning urban resilience analysis is to focus on the methodology required to analyze cities.

2.2 *A systemic approach of cities*

For Pasche and Geisler, the resilience concept comprises individual preventive and emergency measures at building and municipal infrastructure scale and a land use policy to adapt building activities to the risk. Thus, for them, flood preparedness is mainly a matter of flood resilient building and hazard awareness. So, in the current discussion on flood resilient cities a strong emphasis is placed on improving the flood performance of buildings. Yet, the city has to be considered as an entity with different systems and vital functions and not merely as a set of concrete buildings (Anema 2009, Lhomme et al. 2010). Moreover, resilience is focused on system analysis and can be used in a generic systems approach for: ecosystem, social system, socio-economic system… That is why, in the context of urban resilience, cities have to be analyzed in a systemic way.

The concept of city seems perfectly clear to everyone, but defining it is complex. For instance, the definition of a city has been approached in a number of distinct ways. Moreover, still recently, city has been approached by very analytical methods and was characterized by segmented (isolated/disciplinary) studies focus on very particular aspects (transportation, urban planning, urban sociology, environmental aspects ....) with few crossing studies. So, each discipline offers their models and their theories on cities. Yet, for understanding the city as a whole, these analytical approaches are not appropriate, particularly because the models become more and more complex. The systemic approach tends to overcome these segmentations and level of complexity. Indeed, systemic approach proposes a common language for different disciplines and can be considered as a good way to study complex system.

The principle is to considered city like a system, especially like a complex one (such systems are not fully predictable, due to the inherent uncertainty in how these systems evolve). In this framework, a systemic model has been designed. In this model, the city is composed by different elements such as population, companies, public infrastructures, housing and networks (Lhomme et al. 2010). Here networks include all infrastructures and facilities necessary for its operation. These components are supported by the environment and they are organized by governance. Governance refers especially to city administration, region government and state government. Urban systems are not closed and are subject to influences from the external world (systems environment), for example, economic shocks, technological advances, and political changes. The system relations with its environment are characterized by exchanges with other cities (raw materials, manufactured goods ...) and of course the waste produced by activities and population. In this system, it is important to distinguish inputs from outputs because outputs will influence (involve) future inputs (principle of feedback).

3 FOCUS ON TWO MAIN METHODS FOR NETWORKS RESILIENCE ANALYSIS

In order to analyze network resilience, a methodology has been developed based on the study of three capacities: resistance capacity, absorption capacity and recovery capacity (Lhomme et al. 2010). These capacities require to find methods for modeling interdependencies and for assessing redundancy.

3.1 Networks interdependencies

Networks are more and more complex, and more and more interconnected with each other. These interdependencies are the critical points that spread a failure from one network to the other. Some of them are now well identified and managed: the water supplier knows that pumps depend on electricity so that he sets alternative power sources to deal with a disruption in electricity supply. For a critical infrastructure, like networks, getting dysfunctional is a phenomenon that transcends by far the failure of any, even major, single component. The often incomprehensible cause of system crashes stems from the inherent features of the critical infrastructures: they are multicomponent systems, prone to cooperative behaviour, and typically responding in a non-linear fashion to stimuli and perturbations (Rinaldi et al. 2001, Boin & McConnell 2007).

To understand the cascading failures among infrastructure systems under random incidents, manmade attacks and natural hazards, many researchers have proposed different methods for modeling and simulation of interdependent infrastructure systems. These models and methods can broadly be divided into two categories. The first category corresponds to predictive approaches (Johansson & Hassel 2010). Predictive approaches aim at modeling and/or simulating the behavior of a set of interconnected infrastructures in order to, for example, investigate how disturbances cascade between the systems. Notable examples include: Agent Based Methods; Inoperability Input- output Methods; System Dynamics Methods; Network or Graph Based Methods; Data Driven Methods (Ouyang et al. 2011). The second category corresponds to empirical approaches. Empirical approaches aim at studying past events in order to increase the understanding of infrastructure dependencies (Johansson & Hassel 2010).

Existing methods and models address the same issue but from different viewpoints. In fact, the main limit concerning these methods is that they are not exhaustive—all the interdependencies can't be modeling because they are not identified. Indeed, empirical approaches and predictive approaches are generally not combined in the best way. Thus identification and modeling of interdependencies are separated, even if some hybrid approaches has been developed.

Thus, the use of safety methods is proposed in order to model impacts of food hazards on network infrastructures and to study interdependencies between different networks. Based on a functional analysis of the network infrastructures, FMEA (Failure Mode and Effect Analysis) was carried out for each network. Then, using these FMEA, failure scenarios can be designed. Indeed, network systems failure scenarios are designed by linking failure causes to failure modes, and then to failure effects (Lhomme et al. 2010). In this way, the failure mechanisms are modelled as series of functional failures representing the relevant physical processes taking place within the system and leading to loss or deterioration of functions.

The principle of functional modeling is to study the interactions between components of a system and its environment in order to establish a link between the functions failure, their causes and effects (Hartford et al. 2004, Serre et al. 2008). Functional modeling allows better understanding of how the system operates and that's why it allows better understanding of failure mechanisms. Thus it is possible to produce failure scenario and to take into account complex interdependencies thanks to these methods (Lhomme et al. 2010).

FMEA has been chosen because this method allows to be exhaustive concerning identification of failure scenarios. Moreover, this method is combined with others safety methods in order to model interdependencies in a common way (Serre et al. 2008). In this way, identification and modeling are not completely separated.

Using these methods, a general methodology has been designed for studying urban networks disturbances caused by specific hazards. In the same time this methodology has been designed to take into account of interdependencies between networks (Fig. 2). It is a five step approach. Starting from classical internal network analysis—crossing exposure and vulnerability of the networks—damages on the networks are determined.

Thanks to network analysis, disruptions are determined for each network in a second step. Then safety methods are used in order to determine scenarios of disruption between different networks (in order to take into account of interdependencies). The fourth step is to check if these scenarios are real thanks to spatial analysis. To conclude, these new disruptions caused by interdependency require studying again their impact on each network (feedback). So others internal networks studies are required, and the process restart to the step 1 or 2. This iterative process finish when there is no new disruption identified.

3.2 Networks redundancy

Concepts from modern *graph theory* are fundamental to enable measuring of these observable differences in network topology and flow types. A graph is a very simple structure consisting of a set of vertices and a family of lines (possibly oriented), called edges (undirected) or arcs (directed), each of them linking some pair of vertices (Berge 1967). A graph structure can be extended by assigning a weight to each edge of the graph. Weighted graphs are used to represent structures in which connections have some numerical values. For example if a graph represents a road network, the weights could represent the distance or the flow of each road. Thus the study of the graph is not only topologic, but also geometric (distance) and

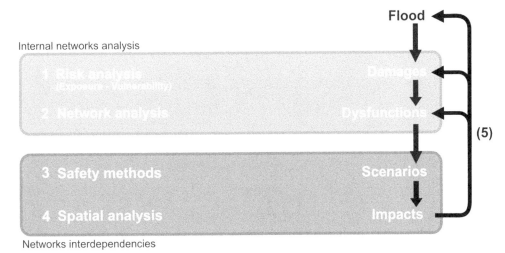

Figure 2. General methodology for studying networks disruptions causing by flood and taking into account of interdependencies between networks.

113

functional (flow). Here we will focus on geometric and topologic properties. Indeed, there is a correlation between the use (the functioning) and the structure of the network (Murray & Grubesic 2007, Winkler et al. 2010). Moreover, studying the underlying network structure of system has proven to be a useful tool, as many features of complex systems arise from their underlying network structure, rather than specifics of the objects, the interactions and the flows (Albert et al. 2004, Winkler et al. 2010).

A network is characterized by a specific capacity to absorb different type of disturbance. Most of the frequent disruptions are always locally absorbed by the networks and the end users remain unaware of their occurrence. This fact results from the ability of the networks to redistribute the flow at the location of the disruption. This is a typical resilience capacity that allows networks operating in a degraded mode. The geometric properties of the network limit the adaptive capacity of the network. Indeed network configuration determines the number of alternative path to disruption of one or several components, in other word the redundancy of the networks.

High network redundancy, requires high opportunities (alternate paths to the shortest path) to get from one point to another point of the graph (Dueñas-Osorio, 2005). Some indicators already exist to quantify the redundancy of a network. For instance, the redundancy ratio assesses the redundancy of a network. This is a good global indicator to characterize redundancy of a network, but it presents some abnormality at a local scale. The clustering coefficient is a measure of the degree to which nodes in a graph tend to cluster together. This indicator is close to the redundancy characterization. Evidence suggests that in most real-world networks, and in particular social networks, nodes tend to create tightly knit groups characterized by a relatively high density of ties (Holland & Leinhardt 1971, Watts & Strogatz 1998). So this indicator doesn't work to characterize a relatively weak density of ties like a tree and a square grid.

Here, redundancy corresponds to the number of "independent" relationships to go from one point to the neighbors of the neighbors of this point. It is proposed for assessing redundancy to count all the independent paths linking one point and the set of neighbors of its neighbors like for example the redundancy ratio (Fig. 3). The difference between this new indicator and the redundancy ratio is that this new indicator is based on the mean of the

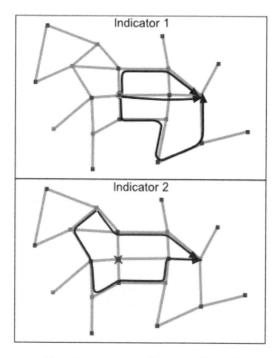

Figure 3.   Two indicators used in order to assess resilience—indicator 1 is based on redundancy ratio and indicator 2 addresses transitivity issue.

114

independent paths (not on the ratio between these independent paths and the total number of independent paths if the graph were complete).

Nevertheless, this indicator is not sufficient to characterize the redundancy of a network. Indeed each point is considered as an origin or a destination point, but in many cases these points are "connection point". The objective is here is to focus analysis on transitivity issues. In mathematics, a binary relation R over a set X is transitive if whenever an element a is related to an element b, and b is in turn related to an element c, then a is also related to c. In our case the relation is not an edge but path which not includes the element b. Thus, redundancy corresponds also to a second indicator corresponding to the number of relation (independent path) between the neighbors of the neighbors of a point when this point is disturbed (Fig. 3).

These two indicators can be aggregated in order to assess redundancy thanks to a single indicator.

4   TOWARD A WEB GIS TOOL IN ORDER TO STUDY URBAN TECHNICAL RESILIENCE

The first methodology presented above allows producing networks failure scenarios taking into account networks interdependencies. Nevertheless, a tool is needed to automate modeling of these scenarios and take advantage of the FMEA completeness. Moreover, redundancy indicators developed require also to be automated. So, these two methods must be automated in a computer tool.

In our work, information about housing, companies, infrastructures, hazards and networks are needed. This type of information is referred to as spatial information, and when visualized, we can see relationships, patterns, and trends that may not otherwise be apparent. A Geographic Information System (GIS) is mapping software that provides spatial information by linking locations with information about that location. It provides the functions and tools needed to efficiently capture, store, manipulate, analyze, and display the information about places and things. That is why we choose to use this type of tool to implement our methods.

It is well known that GIS can be used to recover the spatial component of risk and it is clear that risk assessments have an important spatial component. For instance, to better respond to post disaster activities geographic information system (GIS) technology provides a logical tool for integrating the necessary information and contributing to preparedness, rescue, relief, recovery and reconstruction effort (Gunes and Kovel 2000, Lembo et al. 2008). GIS is seen as a necessary tool in the area of emergency response (Carrara and Guzzetti 1996, Ware 2001). But now resilience requires looking beyond lonely emergency response in order to optimize recovery after a flood event thanks to preparedness and resilience assessment.

That is why a GIS prototype has been developed in order to implement methods. First, this tool will implement the urban systemic model for the data integration and scenarios. Secondly, the tool will implement the redundancy indicators and will be able to take into account of the interdependencies between networks. For instance, two new indicators are implemented and are used for assessing the redundancy of networks. Algorithms are carried out on different types of graph.

This graph represents the road network of the Agglomeration of Orleans (Fig. 4). We will focus on two cities: Bou (left) and Chanteau (right). They are two little cities close to Orleans in the department of Loiret in France and present very different characteristics in terms of redundancy. Indicators have been calculated and a classification in four different classes has been carried out for each indicator. Thus two different maps have been obtained. The last step consists in aggregating information in a single map. For this, for each node, the less redundant class has been chosen to characterize redundancy. Thus, the two maps below show the difference of redundancy between the two cities in a very simple way (Fig. 4). It is important to understand that the agglomeration is composed by thousands of nodes. So, it is impossible to detect easily the difference in terms of redundancy between two cities without these indicators and the prototype. At the end, results emphasize the need to have specific strategies to manage flood risk in the city of Chanteau.

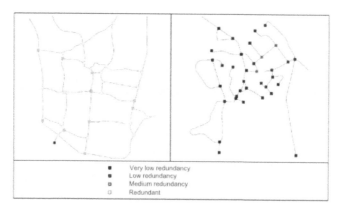

Figure 4. Use of GIS prototype to study networks redundancy—Application on the agglomeration of Orleans (left the city of Bou and right the city of Chanteau).

5 CONCLUSION

This paper proposes methods to improve resilience using spatial analysis. In our methodology, the systemic approach of the city emphasizes the importance of the networks concerning urban resilience. Principles and assumptions finally lead us to analyze how urban networks are able to face natural hazards. In particularly, the methodology highlights the need for a good modeling of networks interdependencies and usefulness of redundancy indicators.

The first prototype used here for the application need again some developments to be fully operational. Nevertheless, at this time, it already produces some results concerning road network redundancy assessment of a French agglomeration. This tool is the first step toward the development of a web GIS studying urban resilience thanks to a better knowledge of networks resilience. This web GIS will be the basis for the negotiation between networks managers and cities planners in order to contribute to build more resilient cities.

To conclude, this research contributes to several improvements in generally separated field of research: urban resilience epistemology with a focus on systemic modeling of cities; graph theory with the development of new redundancy indicators; interdependencies modeling between critical infrastructures with a methodology based on safety methods; urban resilience implementation with the development of specific Web-GIS.

REFERENCES

Albert, R., Albert, I., Nakarado, G., 2004. Structural vulnerability of the North American power grid. *Physical Review E*, 69(2), pp. 1–4.
Anema, K.A., 2009. A scoring system for the resilience of critical infrastructure and their interdependent networks in cities, *International Conference on Urban Flood Management*, Road map towards a flood resilient urban environment Session, UNESCO, Paris, 7 p.
Aschan-Leygonie, C., 2000. Vers une analyse de la résilience des systèmes spatiaux, *L'Espace Géographique*, n°1, p. 67–77.
Ashley, R., Blanksby, J., Chapman, J., Zhou J.J., 2007. Towards Integrated Approaches to Reduced Flood Risk in Urban Areas. In: Ashley, R., Garvin, S., Pasche, E., Vassilopoulos, A., Zevenbergen, C. *Advances in Urban Flood Management*, Taylor & Francis Group, pp. 415–432.
Berge, C., 1967. *Théorie des graphes et ses applications*, Dunod, Paris, 267 p.
Blancher, P., 1998. Risques et réseaux techniques urbains, Centre d'études sur les réseaux, les transports, l'urbanisme et les constructions publiques (CERTU).
Boin, A., McConnell, A., 2007. Preparing for Critical Infrastructure Breakdowns: The Limits of Crisis Management and the Need for Resilience. *Journal of Contingencies and Crisis Management*, 15(1), pp. 50–59.

Campanella, T.J., 2006. Urban Resilience and the Recovery of New Orleans, *Journal of the American Planning Association*, pp. 141–146.

Carrara, A., Guzzetti, F., 1996. Geographical information systems in assessing natural hazards, Advances in Natural and Technological Hazards Research, Vol. 5, springer.

Chen, A., Yang, H., Hong, K.L., Wilson, H.T., 2002. Capacity reliability of a road network: an assessment methodology and numerical results, *Transportation research*, pp. 225–252.

Comfort, L.K., Boin, A., Demchak, C., 2010, *Designing Resilience. Preparing fo Extrem Events*, Pittsburgh, University of Pittsburgh Press.

Demšar, U., Spatenková, O., Virrantaus, K., 2008. Identifying critical locations in a spatial network with graph theory. *Transactions in GIS*, 12(1), Wiley Online Library, pp. 61–82.

Dueñas-Osorio, L., 2005. Interdependent response of networked systems to natural hazards and Intentional disruptions, *Dissertation, Georgia Institute of Technology*. Atlanta: Georgia TechLibrary, UMI 3198529.

Electricity Consumers Resource Council (ELCON). 2004. The economic impacts of the August2003 blackout. http://www.elcon.org/Documents/EconomicImpactsOfAugust2003Blackout.pdf (accessed September 22, 2008).

Folke C., 2006. Resilience: The emergence of a perspective for social–ecological systems analyses, *Global Environmental Change*, pp. 253–267.

Johansson, J., Hassel, H., 2010. An approach for modelling interdependent infrastructures in the context of vulnerability analysis. *Reliability Engineering & System Safety*, 95(12), Elsevier, pp. 1335–1344.

Gallopin C.G., 2006. Linkages between vulnerability, resilience, and adaptive capacity, *Global Environmental Change* 16, pp. 293–303.

Gunderson L.H., Holling C.S., 2002. *Panarchy: Understanding Transformations in Human and Natural Systems*, Island Press, 508 p.

Gunes, A.E., Kovel, J.P., 2000. Using GIS in emergency management operations, *Journal of Urban Planning Development*, 126(3), pp. 126–149.

Hartford., D.N.D., Beacher., G.B., 2004. *Risk and uncertainty in dam safety*, CEA Technologies Dam Safety Interest Group, London, Thomas Telford Eds.

Holland P., Leinhard S., 1971, *Structural sociometry, Perspectives on Social Network Research*, edited by P. Holland and S. Leinhardt, Academic Press, New York.

Holling, C.S., 1973. Resilience and stability of ecological systems, *Annual Review of Ecological System*, pp. 1–23.

Klein, R.J.T., Nicholls, R.J., Thomalla, F., 2003. *Resilience to natural hazards: How useful is this concept?* Environmental Hazards 5 (2003), pp. 35–45.

Lembo Jr A., Bonneau A., O'Rourke T., 2008, Integrative technologies in support of GIS-based post-disaster response, *Natural Hazards Review*, ASCE, pp. 61–69.

Lhomme, S., Serre, D., Diab, Y., Laganier, R., 2010. Les réseaux techniques face aux inondations ou comment définir des indicateurs de performance de ces réseaux pour évaluer la résilience urbaine, *Bulletin de l'association des géographes français*, 2010-n°4, pp. 487–502.

Lhomme, S., Toubin, M., Serre, D., Diab, Y., Laganier, R., 2011. From technical resilience toward urban services resilience, *Proceedings of the fourth Resilience Engineering Symposium*, June 8–10 2011, Sophia Antipolis, France, Presses des Mines, collection sciences économiques, ed. Hollnagel E., Rigaud E., Besnard D., pp. 172–177.

Murray, A.T., Grubesic, T.H., 2011. Critical infrastructure protection: The vulnerability conundrum. *Telematics and Informatics*, 29(1), Elsevier Ltd, pp. 56–65.

Nie, L.M., Lindholm, O., Braskerud, B.C., 2009. Urban flood management in a changing climate, *Journal of Water by Norwegian Water Association*, Vol. 2, pp. 203–213.

Rinaldi, S., Peerenboom, J., Kelly, T., 2001. Identifying, Understanding, and Analyzing Critical Infrastructure Interdependencies, IEEE Control Systems Magazine, IEEE, December 2001, pp. 11–25.

Serre, D., Peyras, L., Tourment, R., Diab, Y., 2008. Levee performance assessment: development of a GIS tool to support planning maintenance actions, *Journal of Infrastructure System*, ASCE, Vol. 14, Issue 3, pp. 201–213.

U.S. Department of Transportation, 2006. Freight on America: a new national picture, http://www.bts.gov/publications/freight_in_america/pdf/entire.pdf

Ware, J.I., 2001. Geospatial data fusion: Training GIS for disaster relief operations, *ESRI proceedings*, 11 p.

Watts, D.J., StrogatzZ, S.H., 1998. Collective dynamics of 'small-world' networks, *Nature, 393*(6684), pp. 409–419.

Winkler, J., Dueñas-Osorio, L., Stein, R., Subramanian, D., 2010. Performance assessment of topologically diverse power systems subjected to hurricane events. *Reliability Engineering & System Safety*, 95(4), pp. 323–336.

Resilience and Urban Risk Management – Serre, Barroca & Laganier (eds)
© 2013 Taylor & Francis Group, London, ISBN 978-0-415-62147-2

Organizational resilience: A multidisciplinary sociotechnical challenge

B. Robert & Y. Hémond
École Polytechnique de Montréal, Montréal, Canada

ABSTRACT: This paper presents the concept of resilience and more specifically the concept of resilience engineering developed in the context of a governmental approach to assess resilience. This approach is part of an overall consequences-based methodology for studying systems in terms of users and providers of resources that aims at identifying the consequences of the outage of a resource on the systems. This approach also allowed including the concepts of dependence and interdependence between systems, which is a key concept in the perspective of evaluating their resilience.

1 INTRODUCTION

In 2005, the United Nations adopted the Hyogo Framework for Action (United Nations, 2005)). This framework has put forward the concept of community resilience and revitalized the research in this area by directly involving the governments. During the same period, some events have highlighted the importance of working on the critical infrastructure protection (CIP). The 9/11 attacks on the World Trade Center, the 2003 blackout in the eastern U.S. and Ontario and the episode of hurricane Katrina in 2005 are only a few of various events that have clearly demonstrated the need for acceptable operation of these infrastructures. These events have, at different levels, generated the failure of various critical infrastructures and have caused negative impacts on multiple critical infrastructures through their interdependencies (Public Safety and Emergency Preparedness Canada, 2006, Prime Minister of Canada, 2011).

CI and their interdependences imply a complex system that needs to be addressed. Resilience and especially resilience engineering can provide concrete results in this direction (Hollnagel, Nemeth and Decker, 2008). The first results of applied work on all critical systems in Quebec have shown the great potential of this approach. These systems, equivalent to CI, interact in a constant changing and interdependent environment.

The object of this paper is to present the theoretical concepts behind resilience engineering and the approaches that relate to it. This concept is currently used in the governmental approach for assessing the resilience of critical systems in Quebec (Organisation de la sécurité civile du Québec [OSCQ], 2009). This approach is part of the global movement to improve the resilience of our societies, in particular that of the CI.

Prior to presenting the concept and its approaches, different definitions and concepts of resilience will be presented. It will then be easier to understand the concept of resilience engineering. Subsequently, the concept of evaluation of resilience will be presented along with the different approaches that support this concept.

2 RESILIENCE: DEFINITIONS AND CONTEXT

Organizational resilience is a concept that has been and still is redefined many times. Present in the literature under two main categories (Dalziel & McManus, 2004), socio-ecological resilience and resilience engineering, it is important to clearly define each one of them. This paper is part of the resilience engineering category (Hollnagel, Nemeth and Decker, 2008; Hollnagel, Woods and Leveson, 2006; Nemeth, Dekker and Hollnagel, 2009). Unlike the socio-ecological resilience (Adger, 2000; Folke, Colding & Berkes, 2003; Holling, 1973; Klein, Nicholls & Thomolla, 2003), resilience engineering focuses on modeling and development of decision-support tools for the industries. Socio-ecological resilience focuses on the study of the systems and their interactions with the environment. Many ideas and concepts are shared by these two categories but the purpose is not necessarily the same. Dalziel & McManus (2004) present the difference between these two categories as follows: resilience engineering concerns the system's performance and its ability to maintain or recover an acceptable level of operation. Socio-ecological resilience concerns the implementation of a system able to provide an acceptable level of operation without searching efficiency in the actions that allow that operation.

The definition used in the context of the governmental approach for assessing the resilience of critical systems in Quebec is the following: "a system's capacity to maintain or restore an acceptable level of functioning despite perturbations or failures" (OSCQ, 2009). This definition integrates various existing definitions while incorporating concepts of engineering of resilience such as the ability to attach to it concrete and operational tools. Three main principles emerge from the definition of resilience:

1. knowledge, the identification of the different components of the system that are required for its proper operation.
2. acceptability, the identification of possible disturbances and their effects on the system and the level of acceptability of these disruptions.
3. adaptability, the identification and evaluation of different tools and measures put in place so the system can adapt to changes in its environment and provide some resilience.

These principles are the keystone of the approach to be presented in the next section.

3 RESILIENCE: EVALUATION AND METHOLODGY

3.1 *Varieties of methodologies*

In each discipline, simple methods to assess the resilience were first developed. Thus, in the field of ecology, Holling (1973) introduced two indicators for evaluating the resilience of an ecosystem:

- the speed of return to equilibrium of a system after a disturbance,
- the amplitude of the disturbance that can be absorbed by the system without changing state.

In the field of economy, at the scale of an enterprise, resilience can be seen as the difference between losses due to the disturbance and the consequences on the company (in percentage). For example, if a 50% loss of electricity causes only a 25% loss of production, resilience will be 25% (50%–25%), (Dauphine & Provitolo, 2007). This type of measure has a limited scope and applicability, since it can only be a post-event evaluation and does not ensure the continuity of the system's functionality.

In the field of risk management, we also find this type of measure. Tierney and Bruneau (2007) propose a representation of resilience in taking into account two parameters: the functionality of the infrastructure and time to return to a normal operation. Here again, we find the concept of return to equilibrium and the resilience is still evaluated following a disaster.

In the UK, resilience is defined as the ability of a system, or organization, to withstand and recover facing adversity. Protection is not the only factor of resilience, it is also supported by effective emergency measures that help reduce the impacts of a failure (Cabinet Office, 2011). In the context of CIs, the UK has recently developed the "Sector resilience plan for critical infrastructure 2011" (Cabinet Office, 2011) in order to evaluate the natural hazards faced by these facilities. The purpose of this plan is to provide a common framework to improve the ability to absorb shocks and to respond quickly to ensure effective emergency measures. The plan must include the four phases of risk management to assess resilience.

The Australian Government has developed a strategy for national resilience (Australian Government, 2010). In this strategy, resilience is defined as the ability of organizations to absorb an event, change, adapt and continue to maintain their competitive advantage and profitability. The strategy aims at evaluating the resilience in terms of continuity of operations by focusing on pre-event preparation and planning rather than on the response and recovery.

In the USA, the National Infrastructure Advisory Council (NIAC) defines resilience in the context of CIs as the ability to reduce the amplitude and/or duration of a disruptive element (National Infrastructure Advisory Council, 2009). At the same time, the NIAC stated that resilience of infrastructure or business depends on its ability to anticipate, absorb, adapt and quickly recover from a disruption, whether the cause is natural or anthropic.

Based on these definitions of NIAC, the Argonne National Laboratory (ANL) has identified three key elements of resilience (Fisher et al., 2010):

- robustness,
- resourcefulness,
- recovery.

The objective of the evaluation of ANL is to obtain an index of resilience for an infrastructure and to compare this index with, other infrastructure in the same sector. It is important to mention the paradigm shift introduced by this model to assess resilience. The evaluation of ANL is not only interested in the various existing plans, it also helps to assess, for example, the level of staff preparedness infrastructure. This gives a relative measure of resilience that takes into account the four pillars of risk management: prevention, preparedness, response, recovery.

In Canada, the government adopted a national strategy on CI (Public Safety Canada, 2010) which gives broad guidelines on the protection of CI and by extension, their resilience. No method for assessing resilience is promoted.

3.2 *Quebec government methodology*

In Quebec, a methodology was developed and applied to 18 ministries and organisms to evaluate their resilience. This methodology (Robert and et al., 2010) includes four main stages, each divided into activities (see table 1).

This methodology focuses on the knowledge of the system in terms of:

1. resources used by the system;
2. resources provided by the system;
3. states of operating of the system (normal, degraded, failure).

Coupled to the consequences-based approach, presented in the next section, it allows characterizing the different consequences generated by the failure of a resource. Subsequently, a characterization of the different mitigation measures (business continuity plans and emergency management plans) allows identifying the various alternatives to ensure resilience of the system, in other words, continuity in the provision of resources.

The result of this characterization is the assessment of resilience. This assessment identifies the parameters related to resilience. This assessment identifies the potential for resilience without concluding that a system is resilient or not. In fact, it is difficult to predict how a system faced with a multitude of perturbations (positive or negative) will respond given the dynamic nature of that system and the environment in which it operates.

This methodology is the result of years of work in the area of interdependences between CI (Robert, de Calan and Morabito, 2008; Robert and Morabito, 2008; Robert and Morabito, 2010a; Robert and Morabito 2010b). Several parameters have been developed to initiate the construction of an overall portrait of the resilience of Quebec's critical systems. These parameters assess the level of expertise of the people/organizations involved in the process, the acceptance of the consequences by the essential resources providers, the coherence of knowledge, the acceptable length of resources outages (disruptions) for the population, the economic activities and the governance. These parameters concern mainly the notion of acceptance and, to a certain extent, the anticipation of disruptions and the planning of effective mitigation measures. Figure 1 shows schematically the initial construction of resilience.

Resilience is a process that must include three parameters: acceptance, anticipation and planning. Since there is no universal measure of the resilience, Figure 1 provides a representation of the state of resilience of an organization based on these three parameters. This organization has made some works with all players on their acceptance of failures and established thresholds. This has enabled the establishment of anticipation measures based on these thresholds. For planning purposes, an inadequate updated of the business continuity plans and emergency measures cause different discontinuity between these points.

The following sections will present the consequences-based approach, approach that is the basis of the characterization and evaluation of resilience. In a second step, the concept of interdependences among critical systems is presented. This concept has helped to adapt the methodology to assess resilience.

Table 1.   Steps of resilience evaluation methodology.

Identification of steps	Description of activities
Step 1 Portrait of the system	Definition of system Identification and break-down of main outputs Identification of functional units
Step 2 Study of outputs and inputs	Characterization of outputs Characterization of inputs Evaluation of conse-quences and response times
Step 3 Management of failures	Identification of critical elements Characterization of man-agement modes Characterization of alterna-tive resources
Step 4 Evaluation of re-silience	Knowledge of system Capacity to maintain its ac-tivities Capacity to restore its ac-tivities State of resilience of sys-tem

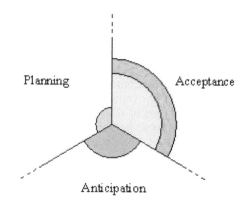

Figure 1.   Illustration of resilience.

4 CONSEQUENCES-BASED APPROACH

The consequences-based approach allows focusing on the various consequences in the event of a resource outage, regardless of the cause that initiated it. This approach has been widely used in the modeling and evaluation of interdependencies among critical systems (Robert and Morabito, 2011) and has yielded concrete results for the different managers involved in the management of these consequences.

This approach implies an understanding of the system as a characterization of the resources used and provided by this system and its various possible states of functioning. It quickly identifies the elements to consider when planning, that is to say when the system is operating normally.

In its current implementation, this approach addresses the risk of cascading failures in critical systems composing the critical infrastructure of a region (Robert et al., 2008; Robert & Morabito 2008, 2010a, 2010b). It is to direct questioning about the consequences that will be generated due to the degradation or loss of a resource known as essential, by taking into consideration two important parameters: time and geographical space. The application of the methodology allows managers and operators of critical systems:

1. to have an overview of the interdependencies;
2. to have an overview or the most sensitive sectors in a region;
3. to anticipate domino effects and the leeway to intervene;
4. to initiate exchanges between the various stakeholders to implement measures of prevention and protection to reduce systems' vulnerabilities and consequences of their failure.

This approach brings the knowledge about the system and its operation that is directly needed to assess the resilience. Modeling and evaluation of interdependencies bring the knowledge required to better understand the functioning of critical systems and to identify inherent vulnerabilities.

The consequences-based approach is the basis of the study of interdependences between critical systems and by extension, to assess resilience. In the next section, the methodology of interdependences will be discussed and a conclusion on the involvement of the latter in the work on engineering of resilience applied to the Quebec government.

5 INTERDEPENDENCES: MODELLING AND ASSESSMENT

Centre risque & performance has developed a methodology to assess interdependencies between CI (Robert and Morabito, 2011). This method allows to model dependencies using an consequence based approach. In this approach, each essential system is considered as an entity that uses resources to provide other resources. Interdependencies between these entities arise from the exchange of these resources (Robert, Morabito and Quenneville, 2007; Robert and Morabito, 2008). When a perturbation affects any of these systems, it can result in a degradation of the resource provided by this system to the society. This degraded resource can then not be used by other systems and can compromise their own functioning (domino effects).

By studying how systems use the resources and how the degradation of a resource can affect them, it is possible to obtain, under the form of a curve, a representation of the state of critical systems as a function of the state of the resources they use (dependencies curves). Figure 2 shows an example of dependencies curves. It presents the dependence of critical systems towards the resource "water" in a certain geographic sector.

From these curves, we find that systems 1 and 2 are not very vulnerable to a water outage. Indeed, although they are affected by this situation, they suffer very small effects that are not likely to jeopardize the achievement of their mission (yellow). Systems 3 and 4, are more vulnerable to a water outage. In their case, they anticipate a loss of their mission after about 4–6 hours, respectively. The periods of 4 to 6 hours are the leeway available to these systems to make decisions and implement alternative measures before their mission is compromised. Finally, the system 5 is extremely vulnerable to a water outage. In this case, the leeway is nil: a loss of water directly impacts the mission of the system 5.

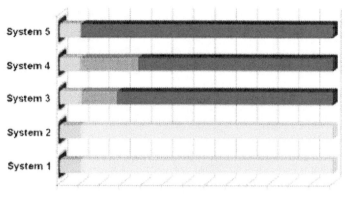

Figure 2. Dependence of systems towards resource water.

This example reveals the extent of application of the consequences-based approach and the simplicity with which it is possible to do so. It is one of the basic concepts essential for assessing the resilience applied to organizations and by extension, engineering of resilience. In the next section, the future prospects of engineering of resilience will be presented highlighting the consequences-based approach and methodology of assessment of interdependences.

6 RESILIENCE ENGINEERING: PERSPECTIVES

Works on the interdependences bring another perspective on the concept of resilience applied to organizations by providing a simple and practical approach to study systems. This approach allows characterizing the resources used, resources available and their interactions within the system. In addition, by focusing on the consequences caused by the failure to provide a resource, it takes into account the interdependencies and the spread of these consequences.

On the other hand, the methodology for assessing the resilience allows the system to engage in a process to ensure its operations and fulfill its mission. This becomes important when it comes to a system (or infrastructure) essential to the functioning of government. The development of simple and concrete tools will ensure continuity in this process.

However, several points remain to be refined. For instance, taking into account the degradation of the resource, rather than a situation "on/off" or the development of the tools mentioned above, this area of research remains to be explored and various projects (Robert and al., 2010; Hollnalgel and al., 2011, Blust, and Lemyre Lemus, 2010) are beginning to show their full potential.

REFERENCES

Adger, W.N. (2000). Social and ecological resilience: are they related? *Progress in Human Geography*, Vol. 24, No. 3, pp. 347–364.
Australian Government. (2010). *Critical Infrastructure Resilience Strategy*. Australia.
Australia Strategic Policy Institute (ASPI). (2008). Strategic Insights 39—Taking a Punch: Building a More Resilient Australia.
Blust, S., Lemus, C. and Lemyre. L. (2010). "Résilience inter-organisationnelle—analyse différentielle de la coordination, coopération et collaboration", *Entretiens Jacques-Cartier 2010*. [Online], http://www.pacte.cnrs.fr/spip.php?article2325 [view May 24th]
Cabinet Office. (2011). UK resilience. Consulté le 17 may 2012, tiré de http://www.cabinetoffice.gov.uk/ukresilience.

Dalziell, E.P., & Mcmanus, S.T. (2004). Resilience, Vulnerability, and Adaptive Capacity—Implications for System Performance. International forum for engineering decision making.

Dauphiné, A., & Provitolo, D. (2007). La résilience: un concept pour la gestion des risques. *Annales de géographie*, 115–125.

Fisher, R.E., Bassett, G.W., Buehring, W.A., Collins, M.J., D.C.;, D., Eaton, L.K., et al. (2010). *Constructing a resilience index for the enhanced critical infrastructure protection program*: Argonne National Laboratory.

Folke, C., Colding, J., & Berkes, F. (2003). Synthesis: Building resilience and adaptive capacity in social-ecological systems. In F. Berkes (Ed.), Navigating social-ecological systems Building resilience for complexity and change (pp. 352–387). Cambridge University Press.

Hémond, Y. and Robert, B. (2011). Evaluation of state of resilience for a critical infrastructure in a context of interdependencies. *Conference proceedings NGINFRA 2011*. IEEE.

Holling, C.S. (1973). Resilience and stability of ecological systems. *Annual review of cology and systemics*, Vol. 4, pp. 1–23.

Hollnagel, E., Nemeth, C.P. and Dekker, S. (2008). *Resilience Engineering Perspectives, Volume 1: Remaining Sensitive to the Possibility of Failure.* Farnham, UK: Ashgate.

Hollnagel, E., Woods, D.D. & Leveson, N. (2006). *Resilience Engineering: Concepts and Precepts.* Farnham, UK: Ashgate.

Hollnagel, E., Pariès, J., Woods, D.D. and Wreathall, J. (2011). *Resilience engineering in practice—a guidebook*, United-States: Ashgate.

Klein, R.J.T., Nicholls, R.J., & Thomalla, F. (2003). Resilience to natural hazards: how useful is this concept? *Global Environmental Change Part B Environmental Hazards*, Vol. 5, No. 1–2, pp. 35–45.

Nemeth, C.P., Hollnagel, E. & Dekker, S. (2009). *Resilience Engineering Perspectives, Volume 2: Preparation and Restoration.* Farnham, UK: Ashgate.

National Infrastructure Advisory Council. (2009). Critical Infrastructure Resilience, Final Report and Recommendations. Consulté le 15 may 2012, tiré de http://www.dhs.gov/xlibrary/assets/niac/niac_critical_infrastructure_resilience.pdf

Organisation de la sécurité civile du Québec (OSCQ). (2009). *Cadre de référence de la démarche gouvernementale de résilience des systèmes essentiels au Québec.* Quebec, QC: OSCQ.

Prime Minister of Canada (2011). Beyond the Border: a shared vision for perimeter security and economic competitiveness. *A declaration by the Prime Minister of Canada and the President of the United States of America.* Online http://www.pm.gc.ca/eng/media.asp?id = 3938. Consulted may 2012.

Public Safety Canada (PSC). (2010). Aller de l'avant avec la Stratégie nationale sur les infrastructures essentielles. Ottawa, Canada: Sécurité publique Canada.

Public Safety and Emergency Preparedness Canada. (2006). *Ontario–U.S. Power Outage—Impacts on Critical Infrastructure.* http://www.publicsafety.gc.ca/prg/em/_fl/ont-uspower-e.pdf

The Reform Institute. (2008). *Building a Resilient Nation: Enhancing Security, Ensuring a Strong Economy.* Washington: The Reform Institute.

Robert, B., de Calan, R. and Morabito, L. (2008) "Modeling interdependencies among critical infrastructures", *Int. J. Critical Infrastructures*, Vol. 4, No. 4, pp. 392–408.

Robert, B. and Morabito, L. (2008) "The operational tools for managing physical interdependencies among critical infrastructures", *Int. J. Critical Infrastructures*, Vol. 4, No. 4, pp. 353–367.

Robert, B. and Morabito, L. (2010a) "An approach to identifying geographic interdependencies among critical infrastructures", *Int. J. Critical Infrastructures*, Vol. 6, No. 1, pp. 17–30.

Robert, B. and Morabito, L., (2010b) "Dependency on electricity and telecommunications", *Securing electricity supply in the cyber age: Exploring the risks of information and communication technology in tomorrow's electricity infrastructure*, Chapter 3, Springer 2010, 187 pages.

Robert, B. and Morabito, L. (2011) *Reducing vulnerability of critical infrastructures—methodological manual.* Montréal: Presses internationalles Polytechnique, 2011.

Robert, B., Morabito, L. and Quenneville, O. (2007) "The preventive approach to risks related to interdependent infrastructures", *Int. J. Emergency Management*, Vol. 4, No. 2, pp. 166–182.

Robert, B. and al. (2010) *Organizational resilience—concepts and evaluation methods*, Montréal: Presses internationalles Polytechnique, 2010.

Tierney, K., & Bruneau, M. (2007). Conceptualizing and measuring resilience: a key to disaster loss reduction. *Transportation Research News, May-June*, 14.

United Nations. (2005). *Report of the World Conference on Disaster Reduction.* World Conference on Disaster Reduction. Kobe, Hyogo, Japan, 18–22 January 2005.

Resilience and Urban Risk Management – Serre, Barroca & Laganier (eds)
© 2013 Taylor & Francis Group, London, ISBN 978-0-415-62147-2

Resilience-based design for urban cities

G.P. Cimellaro & V. Arcidiacono
Department of Structural, Geotechnical and Building Engineering (DISEG),
Politecnico di Torino, Turin, Italy

ABSTRACT: A new design methodology is developed called "Resilience-Based Design" (RBD) which can be considered as an extension of Performance-Based. The goal of RBD is to make individual structures and communities as "Resilient" as possible, developing technologies and actions that allows each structure and/or community to regain its function as promptly as possible. The paper describes a holistic framework for measuring disaster resilience at the community scale. Seven dimensions characterizing community functionality have been represented by the acronym PEOPLES. The proposed Framework provides the basis for development of quantitative models that measure continuously the functionality and resilience of communities against extreme events. In order to show the application issues, the road network of Treasure Island in San Francisco Bay has been considered as case study. Interdependencies among lifelines and the housing facility have been also considered. Over the longer term, this framework will enable the development of geospatial and temporal decision-support software tools that help planners to enhance the resilience of their communities.

1 INTRODUCTION

Over the last two decades the frequency and severity of natural disasters increased significantly and this trend is expected to continue. The 2004 Tsunami in South Asia, the 2005 Hurricane Katrina in US, the 2009 Great Britain and Ireland floods, the 2009 Italian earthquake, the 2010 Volcanic eruption of the Eyjafjallajökull in Iceland, the 2010 Central Europe floods and the 2011 Tohoku Earthquake in Japan are only few dramatic examples of the last decade. These events have shown how systems (regions, communities, cities, structures, etc.) are vulnerable to natural hazards resulting in devastating disasters as it is also the case after occurrence of human errors, systems failures, pandemic diseases etc. As the world become wealthier and wealthier, the megacities start increasing so fast, but no city so far has demonstrated to build a resilient city. The main question is how can we reduce the risk in urban cities? Shall we move all the entire population from the risky areas (e.g. away from the coast, the earthquake faults, volcano etc.)? Provided this is not possible, because the risky areas are usually the most beautiful areas of our planet, then we should realize preventive measures which will save money and lives. Disaster should be prevented to happen and indeed when it does happen, its effects should be reduced. In order to reduce the socio-economic losses and environmental consequences that affect the whole society, the emphasis has shifted to mitigation and preventive actions to be taken before the extreme event might happen. Mitigation actions can reduce the vulnerability of a system and reduce the recovery time required to regain stability and functionality of the community. Therefore, there is also a need for cost-effective mitigation of potential and actual damage from disruptions, particularly those causing cascading effects capable of incapacitating a system or an entire region and of impeding rapid response and recovery. To evaluate the capacity of a community to cope with and manage a catastrophic event, it is necessary to provide quantitative and objective evaluation metrics of its capacity to respond for each term. The most effective way to protect and prepare a community is by evaluating its ability to deal with and recover from a disaster.

While studies on the disaster resilience of technical systems have been undertaken for quite some time (Chang and Shinozuka, 2004), the societal aspects and the inclusion of various and multiple types of extreme events are new developments. In this regard, communities around the world are increasingly debating ways to enhance their resilience.

In earlier work by Bruneau et al. (2003), resilience was defined including technical, organizational, economic, and social aspects and with four main properties of robustness, redundancy, resourcefulness, and rapidity. The quantification and evaluation of disaster resilience was based on non-dimensional analytical functions related to the variations of functionality during a "period of interest," including the losses in the disaster and the recovery path (Cimellaro et al., 2009). This evolution over time, including recovery, differentiates the resilience approach from other approaches addressing only loss estimation and its momentary effects.

The objective of this paper is to describe a holistic framework for defining and measuring disaster resilience for a community at various scales. Seven dimensions characterizing community functionality have been identified and are represented by the acronym PEOPLES (Renschler et al., 2010): Population and Demographics, Environmental/Ecosystem, Organized Governmental Services, Physical Infrastructure, Lifestyle and Community Competence, Economic Development, and Social-Cultural Capital. The proposed PEOPLES Resilience Framework provides the basis for development of quantitative and qualitative models that measure continuously the functionality and resilience of communities against extreme events or disasters in any or a combination of the above mentioned dimensions.

A new more general design methodology is developed in this paper called "Resilience-Based Design" (RBD) which can be considered as an extension of Performance-Based Design which is only a part of the total "design effort". The goal of RBD is to make individual structures and communities as "Resilient" as possible, developing technologies and actions that allows each structure and/or community to regain its function as promptly as possible. Over the longer term, this framework will enable the development of geospatial and temporal decision-support software tools that help planners and other key decision makers and stakeholders to assess and enhance the resilience of their communities.

2 DEVELOPMENT OF PBEE

In order to understand RBD, this paragraph summarizes in the bullet list below the steps of Performance-Based design in US in the last two decades:

- 1990s, Performance-based earthquake engineering;
- 1997, First-generation of Performance-based Earthquake Engineering (FEMA 273: NEHRP Guidelines for the Seismic Rehabilitation of Buildings);
- 2000, Second-generation Performance-Based Earthquake Engineering (FEMA 356: Prestandard and Commentary for the Seismic Rehabilitation of Buildings);
- 2006, Next generation Performance-based Earthquake Engineering (FEMA 445: Next generation Performance-Based Seismic Design guidelines);

The main equation of PBD is based on the total probability theorem de-aggregating the problem into several interim probabilistic models (namely seismic hazard, demand, capacity and loss models) and it is defined as follows (Cornell and Krawinkler, 2000).

$$\lambda\left(dv < DV\right) = \underbrace{\iiint \underbrace{G\left(dv|dm\right)}_{\text{Loss Analysis}} \underbrace{dG\left(dm|edp\right)}_{\text{Damage Analysis}} \underbrace{dG\left(edp|im\right)}_{\text{Response Analysis}} \underbrace{\left|d\lambda\left(im\right)\right|}_{\text{PSHA}}}_{\text{Seismic risk}} \tag{1}$$

where im = intensity measure (e.g $Sa(T_1)$, epsilon, $S_{dinelastic}$, duration etc.); dm = damage measure (e.g physical condition & consequences/ramifications); edp = engineering demand parameters (e.g. drift ratio (peak, residual), acceleration, local indices etc.); dv = decision variable (e.g. loss, functionality, downtime, casualties etc.); $\lambda(dv)$ = mean annual frequency of a decision variable (dv); $G(a|b)$ is the probability of exceedance $a > a_0$ given b. Each component of

Eq. (1) need to be determined statistically. Currently the method has been implemented in ATC-58 and ATC-63, but also in other countries like in China where PBEE has been added to the new version of the code "Seismic Design for Building Structures" (GB50011–2010), to design tall buildings and innovative systems, while majority of buildings could be still designed with traditional RSA.

2.1 *Limitations of PBEE*

Although this methodology is rapidly spreading, there are parts that the PBEE does not cover such as:

1. The portfolio assessment;
2. Community assessment.

The concept of Performance-Based Design/Engineering can be applied to describe the behaviour of a single building or structure, but the performance of an individual structure is not governed by its own performance, but interacts heavily with the performance of other entities within the same community. A clear example of these interdependencies between the building and the community is for example a hospital. Its functionality cannot be considered independent of the rest of the system (e.g. a hospital unit without water and electricity or without roads connecting it to the rest of the world cannot be considered functional even if it does not have any structural damage). This example in very simple terms distinguishes between PBD which would consider the house ok, while in RBD the house would not satisfy the requirements. Another example of the limitations of PBD is given by 2009 L'Aquila earthquake (Cimellaro et al., 2010c), during which the small town of Castelnuovo was completely destroyed, except a single housing unit that was standing after the earthquake and suffered minor damage. According to PBD the building is ok, but in RBD the housing unit would be not operational, because it is not able to interact with other entities inside the same community.

3 RESILIENCE-BASED DESIGN

Today, designers and engineers approach a structure as if it stands alone, without considering the interaction with the community, which instead should be considered as an integrated part of the design process. There is now a new fundamental way of looking at all the problem. The building is not considered alone, but as a group of buildings using the "Portfolio Approach" which will allow regional loss analysis. So it will be moved from housing units to housing blocks.

This concept is borrowed from the financial industry, where Modern Portfolio Theory (MPT) was developed in the 1950s through the early 1970s and was considered an important advance in the mathematical modelling of finance. MPT is defined as a theory of investment which attempts to minimize risk for a given level of expected return (performance), by carefully choosing the proportions of various assets. MPT is a mathematical formulation of the concept of diversification in investing, with the aim of selecting a collection of investment assets that has collectively lower risk than any individual asset.

Analogously the concept of diversification can be applied in the field of disaster resilience, where diversification in retrofit of different buildings in a given region can increase resilience collectively more than any individual retrofit.

MPT models an asset's return as a normally distributed function, defines risk as the standard deviation of return, and models a portfolio as a weighted combination of assets so that the return of a portfolio is the weighted combination of the assets' returns. Similarly the risk in RBD can be defined as the standard deviation of the performances of each housing unit; therefore the performances of a given community can be defined as the weight combination of the performance of each housing unit. It is important to mention that community is a complex system where losses as well as the recovery process are coupled dimensions which involve several parameters which are not only engineering parameters such as drift and accelerations, but also other parameters such as socio-economic gender, age of the population etc. All these parameters are used to define the recovery that becomes part of the design process in RBD and therefore should be planned upfront.

3.1 *Definition of resilience*

At this time, there is no explicit set of procedures in the existing literature that suggests how to quantify resilience in the context of multiple hazards, how to compare communities with one another in terms of their resilience, or how to determine whether individual communities are moving in the direction of becoming more resilient in the face of various hazards. Considerable research has been accomplished to assess direct and indirect losses attributable to earthquakes, and to estimate the reduction of these losses as a result of specific actions, policies, or scenarios. However, the notion of resilience suggests a much broader framework than the reduction of monetary losses alone.

Resilience (R) may be defined as a function indicating the capability to sustain a level of functionality or performance for a given building, bridge, lifeline networks, or community, over a period defined as the control time T_{LC} that is usually decided by owners, or society at large, for example, and corresponds to the expected life cycle or life span of the building or other system. Resilience, R, is defined graphically as the normalized shaded area underneath the function describing the functionality of a system, defined as $Q(t)$. $Q(t)$ is a non-stationary stochastic process, and each ensemble is a piecewise continuous function as shown in Figure 2, where $Q(t)$ is the functionality of the region considered. The change in functionality due to extreme events is characterized by a drop, representing a loss of functionality, and a recovery.

Analytically Resilience is defined as

$$R(\vec{r}) = \int_{t_{OE}}^{t_{OE}+T_{LC}} Q_{TOT}(t)/T_{LC}\, dt \tag{2}$$

where $Q_{TOT}(t)$ is the global functionality of the region considered (Cimellaro et al. 2005, 2006) which is given in Eq. (3). The functionality is the combination of all functionalities related to different facilities, lifelines, etc. for the case when physical infrastructures resources and services are considered which will be described in the following paragraph.

This selection will define the performance measures to be considered to define the global functionality of the system (Eq. (3)).

Resilience can be considered as a dynamic quantity that changes over time and across space. It can be applied to engineering, economic, social, and institutional infrastructures, and it can use various geographic scales (Figure 3).

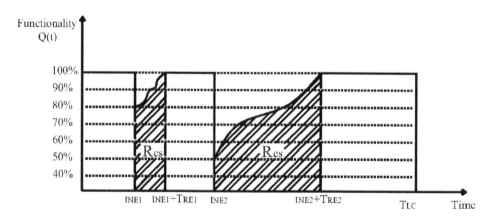

Figure 1. Community resilience.

130

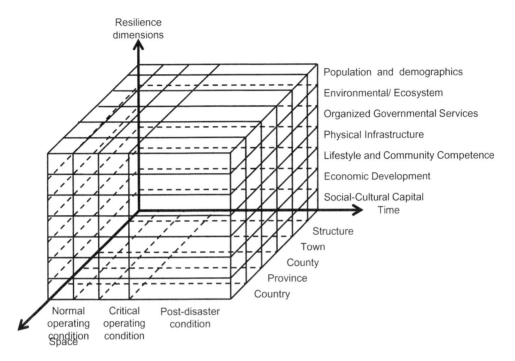

Figure 2. Spatial and temporal dimension of Resilience-Based Design (RBD) using PEOPLES approach.

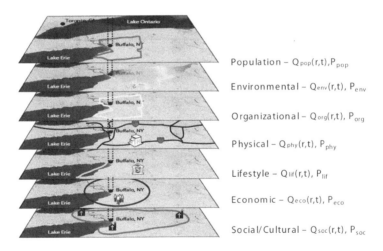

Population – $Q_{pop}(r,t)$, P_{pop}

Environmental – $Q_{env}(r,t)$, P_{env}

Organizational – $Q_{org}(r,t)$, P_{org}

Physical – $Q_{phy}(r,t)$, P_{phy}

Lifestyle – $Q_{lif}(r,t)$, P_{lif}

Economic – $Q_{eco}(r,t)$, P_{eco}

Social/Cultural – $Q_{soc}(r,t)$, P_{soc}

Figure 3. Resilience index using PEOPLES methodology.

The first step in order to quantify the resilience performance index (R) is to define the spatial scale (e.g. building, structure, community, city, region. etc.) of the problem of interest (Figure 3).

The second step is to define the temporal scale (short term emergency response, long term re-construction phase, midterm reconstruction phase etc.) of the problem of interest (Figure 3). The selection of the control period T_{LC} will affect the resilience performance index, therefore when comparing different scenarios the same control period should be considered.

Details about the description of each one of the dimensions of the PEOPLES framework can be found in Renschler et al. (2010a, 2010b), but in this paper are described only the physical infrastructure and economic dimensions that are summarized in the next two subparagraphs.

3.2 The seven dimensions of community resilience

Disaster resilience is often divided between technological units and social systems. On a small scale, when considering critical infrastructures, the focus is mainly on technological aspects. On a greater scale, when considering an entire community, the focus is broadened to include the interplay of multiple systems—human, environmental, and others—which together add up to ensure the functioning of a society. In order to emphasize the primary role of the human system in community sustainability, the acronym "PEOPLES" (Rencshler et al. 2010) has been adopted in order to describe the framework that is built on and expands previous research at MCEER linking several previously identified resilience dimensions (technical, organizational, societal, and economic) and resilience properties (r4: robustness, redundancy, resourcefulness, and rapidity) (Bruneau et al. 2003, Cimellaro et al. 2009). PEOPLES incorporates MCEER's widely accepted definitions of service functionality, its components (assets, services, demographics) and the parameters influencing their integrity and resilience.

3.3 Physical infrastructures

In order to clarify the concept let's focus on one of the dimensions of the PEOPLES framework. The physical infrastructure dimension incorporates both facilities and lifelines. Within the category of facilities, residential facilities, commercial facilities, and cultural facilities are included. Lifelines are those essential utility and transportation systems that serve communities across all jurisdictions. Lifelines are thus components of the nation's critical infrastructure, which also includes medical, financial, and other infrastructure systems that create the fabric of modern society.

Without water and electricity, critical facilities such as hospitals cannot perform effectively their primary functions. Inaccessible roads make surface transportation impossible, creating an obstacle for supply chain management and efficient movement. When streets

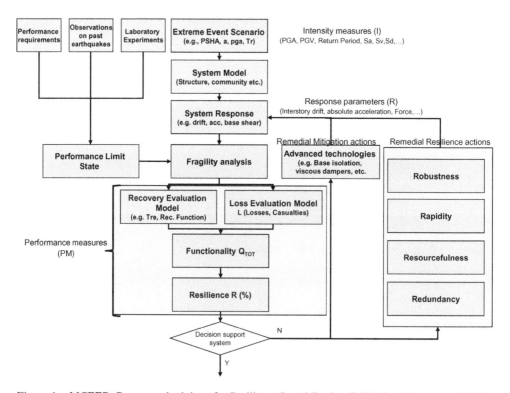

Figure 4. MCEER Center methodology for Resilience-Based Design (RBD) based on control (feedback loop) approach.

and buildings are cordoned off because of damage, businesses may be open, but will not be "in business." Even when businesses relocate for the short-term due to damage to facilities, customers may not find the businesses. In terms of housing, key indicators may include proportion of housing stock not rated as substandard or hazardous and vacancy rates for rental housing (Tierney 2009). In terms of communication networks, key indicators may include adequacy (or sufficiency) of procedures for communicating with the public and addressing the public's need for accurate information following disasters, adequacy of linkages between official and unofficial information sources, and adequacy of ties between emergency management entities and mass media serving diverse populations. In the aftermath of a disaster, the restoration and recovery of physical infrastructure remain by-and-large technical issues. However, those are tightly related and often driven by organizations, economics and socio-political events. Resilience must consider these interactive dimensions in order to be relevant to the system.

3.4 Layer's model

The general framework at the community level is described by the following equations where for each dimension a performance indicator and/or functionality is defined which is combined with other functionality dimensions as follows

$$Q_{TOT}(t) = Q_{TOT}(Q_P, Q_E, Q_O, Q_{Ph}, Q_L, Q_E, Q_S) \tag{3}$$

where Q_{TOT} = global functionality; Q_X = functionality of each of the seven dimensions according to PEOPLES framework. Within each dimension functionality is defined as a combination of functionality of the respective subsystems, for example the functionality of the physical infrastructures is defined as follows

$$Q_{Ph}(t) = Q_{Ph}(Q_{Hosp}, Q_{Ele}, Q_{Road}, Q_{Water}, \ldots) \tag{4}$$

This list of functionality terms that can be inserted within the physical infrastructure block, but it is not complete and additional terms can be eventually added, such as functionality of schools, stations, fire stations, oil and natural gas systems, nuclear facilities, emergency centers, communication towers/antennae etc. Once the geographic scale and the global functionality Q_{TOT} is defined it is possible to plot its value over the region of interest in a contour plot at a given instant of time t, so time-dependent functionality maps of the region can be obtained. When also the temporal scale is defined through the control time T_{LC}, then the resilience contour map of the region of interest can be plotted. The Resilience maps will vary in space from point to point, but it will be time independent and described by Equation (2).

		Earthquake Performance Level			
		Fully Operational	Operational	Life Safe	Near Collapse
Earthquake Design Level	Frequent (43 years)	Basic Objective	Unacceptable	Unacceptable	Unacceptable
	Occasional (72 years)	Essential/Hazardous Objective	Basic Objective	Unacceptable	Unacceptable
	Rare (475 years)	Safety Critical Objective	Essential/Hazardous Objective	Basic Objective	Unacceptable
	Very Rare (975 years)	Not Feasible	Safety Critical Objective	Essential/Hazardous Objective	Basic Objective

Figure 5. Recommended seismic performance objectives for buildings (SEAOC Vision, 2000).

133

Finally the community resilience index is given by the double integral over time and space as follow

$$R_{com} = \int_{A_C} R(\vec{r}) A_C dr = \int_{A_C} \int_{t_{OE}}^{t_{OE}+T_{LC}} Q_{TOT}(t)/(A_C T_{LC}) dt dr \qquad (5)$$

For each dimension a contour plot can be determined and combined using a layered approach as the one shown in Fig. 3. Then a radar graph can be plotted and the area will define the final value of the resilience score for the region of interest.

In summary a schematic step-by-step procedure of the MCEER methodology described in Figure 6 is the following:

1. Define extreme event scenarios (e.g. PSHA, ground motion selection);
2. Define the system model;

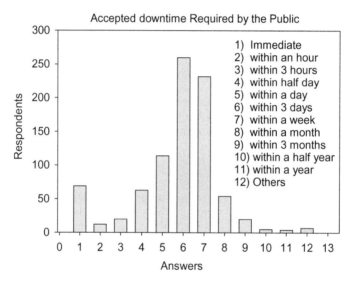

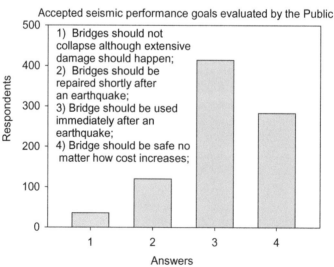

Figure 6. (a) Accepted downtime required by the public; (b) Accepted seismic performance goals evaluated by the public (Yokohama National University, 2011).

3. Evaluate the response of the model;
4. Compute different performance measures (e.g. losses, recovery time, functionality, resilience);
5. Identify remedial mitigation actions (e.g. advanced technologies) and/or resilience actions (e.g. resourcefulness, redundancy, etc.).

This design approach has analogies with the feedback loop taken from control theory.

3.5 *Uncertainties in Resilience-Based Design*

Either a deterministic or probabilistic approach can be used within the RBD with preference to the latter approach when a particular level of confidence of achieving performance objective is of interest. The MCEER methodology also has a probabilistic approach which is more general even if the information provided to the public (e.g. decision makers, politicians, etc.) should be deterministic, because it is more simple and easy to understand. The main equation of RBD based on the total probability theorem is as follows (Cimellaro et al., 2006, 2010)

$$\bar{R} = \int\limits_{T_{RE}} \int\limits_{L} \int\limits_{DM} \int\limits_{R} \int\limits_{i^*} r_i \cdot P(T_{RE}/Q)P(Q/PM)P(PM/R)P(R/I)P(I_{T_{LC}} > i^*)dI \cdot dR \cdot dPM \cdot dQ \cdot dT_{RE}$$

(6)

4 DEFINITION OF RESILIENCE PERFORMANCE LEVELS

The objective of Performance Based Seismic Engineering (PBSE) is to design, construct and maintain facilities with better damage control. A comprehensive document has been prepared by the SEAOC Vision 2000 Committee (1995) that includes interim recommendations. The performance design objectives couple expected or desired performance levels with levels of seismic hazard as illustrated by the Performance Design Objective Matrix shown in Figure 7.

Performance-Based design levels focus on the performances a building can hold during the shaking and are associated to engineering demand parameters such as deformations. More recently SPUR (SPUR, 2009) which is the San Francisco planning and Urban Research Association introduced other definitions of performance levels for physical infrastructures based on recovery target states which take in account the safety as well as the recovery time. Five performance measures for buildings have been identified:

1. Safe and Operational;
2. Safe and usable during repair;
3. Safe and usable after repair;
4. Safe but not repairable;
5. Unsafe.

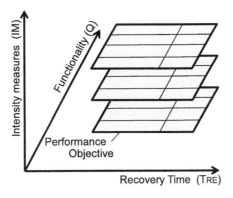

Figure 7. 3-dimensional Resilience Performance objectives matrix for structures, communities, systems etc.

In Japan, there was a trial to find the desired performance goals in downtime and bridge performance from public interview. This information will be useful guideline to define the boundaries of the different resilience performance levels.

The new proposed Resilience Performance Levels (RPL) focus on building performance after the earthquake stops, recognizing the importance of the temporal dimension (Recovery time T_{RE}) in the assessment of the RPLs of structures and communities in general. A 2-D performance domain consisting of Performance Levels PL(i, j), defined by the combination of Functionality (index j) and recovery time (index i) is proposed. By accounting for the effect of the temporal dimension, a 3-dimensional performance matrix (Figure 7) can be visualized as a set of predefined joined performance domains ("masks") for different seismic intensity level, IM and different RPLs.

		Recovery Time		
		Short term (Emergency)	Midterm	Longterm (Reconstruction)
Functionality Performance Levels	Fully Operational (Q1)	Basic Objective	Unacceptable	Unacceptable
	Operational (Q2)	Essential Objective	Basic Objective	Unacceptable
	Partially Operational (Q3)	Critical Objective	Essential Objective	Basic Objective
	Near not Operational (Q4)	Not feasible	Critical Objective	Basic Objective

5 EXAMPLES OF APPLICATION OF RESILIENCE-BASED DESIGN

5.1 *Hospital system*

One of the requirements of RBD is to achieve redundancy in a given community which is a complex system of interacting sub-systems where you can measure both the performance objectives of individual structures and the performance objectives of the entire community. In a selected region resilience can be improved by increasing the number of hospitals in the region advocating redundancy. However, one of the aspects that should be taken in account in RBD which is not taken in account in the current practice is that for example if the region is affected by the same earthquake hazard is it resilient to build the new hospital with the same retrofit technique of the already existing hospital? Selecting different retrofit strategies for the new hospital might enhance the probability to survive in case of extreme event of at least one hospital. Following the same concept an example of design that has followed the current practice of PBD is the C.A.S.E. project realized after 2009 L'Aquila earthquake in Italy. It allowed realizing several new housing units with the same type of retrofit that consists in a base isolation system realized with friction pendulum bearings. If, in a very rare scenario, the next earthquake would be characterized by low frequency content, probably all these new housing units would collapse, because they do not have diversity in performance. When looking at all housing units as a system, they are not redundant, because Resilience has not been considered in the design process.

An example based on a series of hospital buildings described by Park et al. (2004) is chosen to illustrate the concept of RBD. They consist of five concrete shear wall systems and one unreinforced masonry bearing system. Four different retrofit actions have been considered from No Action to the Rebuild Option. If uncertainties in the seismic input are considered by using four different hazard levels, then resilience index can be evaluated using Equation (2) for different rehabilitation strategies and compared as shown in Figure 11. For this case study it is shown that the Rebuild option has the largest disaster resilience of 98.7%, when compared with the other three strategies, but it is also the most expensive solution ($ 92.3 millions).

However, if No Action is taken the disaster resilience is still reasonably high (65.0%). Further detail about this case study can be found in Cimellaro et al. (2010).

5.2 *Road network of Treasure Island in San Francisco Bay*

Twenty-five buildings located in Treasure Island in San Francisco bay have been selected as case study (Fig. 9). The island is connected to San Francisco and Oakland through the Bay Bridge which is located on highway 80.

It is assumed an earthquake with a return period of $T_r = 2450\ yrs$. The secondary roads have been modelled with an "equivalent road segment" with a weight equivalent to the total length of the secondary roads which are located in the area of influence of the link (Fig. 10a). In this case study, interdependencies between the road network and the building unit has been taken in account. When a building collapses in the area of influence near a road, the link will lose its functionality (Fig. 10). The area of influence is rhomboidal if the road is secondary (Fig. 10b), while it is rectangular if it is a main road (Fig. 10c).

The main assumptions of the recovery scenario are: (i) no-limit on the economic budget (EB); (ii) maximum of 3 simultaneous starts of four construction building sites (CSS); and also (iii) maximum of 3 construction sites per day inside the selected area (CS). In Fig. 11 it is possible to see the road network and the housing units that have been modelled.

In Fig. 11 in red are the roads which are not accessible after the earthquake, while in blue are the roads which are still accessible. The functionality curves of the housing units and the road network are plotted in Fig. 12. Equal weight has been given to the road network and the housing units; therefore the total functionality is simply the sum of the two functionalities.

The functionality curve of the road network at 40th day has a leap due to the recovery of the first bridge that joints the Island with the mainland (see Fig. 12b). The final resilience

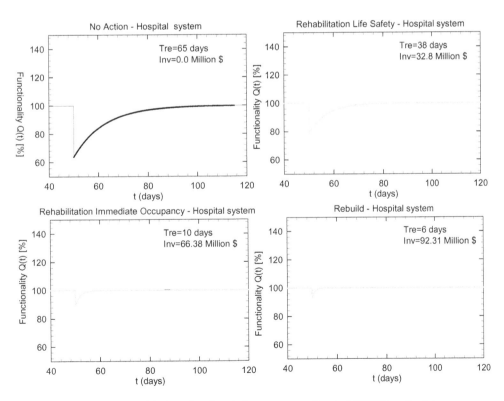

Figure 8.   Resilience-based design of an hospital system according to MCEER methodology.

Figure 9.   Treasure Island in San Francisco Bay.

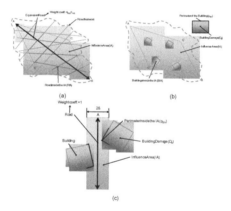

Figure 10.   Equivalent road network concept and interdependencies with housing units.

Figure 11.   Road network modeling in the Bay Area.

138

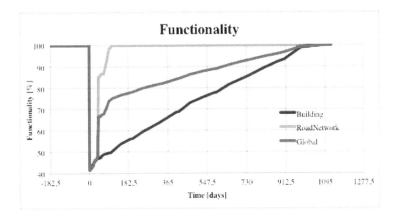

Figure 12. (a) Functionality curves; (b) Functionality after the recovery of the first bridge in the Bay Area.

indices are 76.20%, 97.44% respectively for buildings, road network, while the total resilience index combining both the buildings and the road network is 86.82% when equal weight is given to both systems.

6 CONCLUSIONS

Climate change is candidate to become the most spectacular, but also the more demanding challenge of the emergencies authorities, at local, regional and national level. Nowadays, the emergency response rely on national authorities, however, disaster prevention and resilience have a trans-border, trans-regional and trans-national dimensions. Therefore the international community becomes conscious that there is need for collaboration and coordination among different states in Europe. A new design methodology is developed in this paper called "Resilience-Based Design" (RBD) which can be considered as an extension of PBD which is only a part of the total "design effort". The goal of RBD is to make individual structures and communities as "Resilient" as possible, developing technologies and actions that allows each structure and/or community to regain its function as promptly as possible.

The goal of this paper is to introduce the concept of Resilience-Based Design and provide a quantitative definition of resilience in a rational way through the use of an analytical function that may fit both technical and organizational issues. The fundamental concepts of community resilience are analyzed and a common frame of reference is established. However, even if the performance evaluation of an individual structure is the engineers' goal, the level of performance required should not be determined individually, but on a community basis. A Resilience Performance level matrix combining functionality and recovery time as performance levels and performance objectives at increasing level of seismic intensity has been presented. The resilience-based approach is illustrated in detail in some examples where interdependencies between lifelines and facilities are taken in account, showing its efficiency and better description with respect to PBD.

ACKNOWLEDGMENT

The research leading to these results has received funding from the European Community's Seventh Framework Programme—Marie Curie International Reintegration Actions—FP7/2007-2013 under the Grant Agreement n° PIRG06-GA-2009–256316 of the project ICRED—Integrated European Disaster Community Resilience.

REFERENCES

Alesch, D. 2005 Complex urban systems and extreme events: toward a theory of disaster recovery. *Proceedings of the 1st International Conference on Urban Disaster Reduction*, Kobe, Japan.

Arcidiacono, V., Cimellaro, G.P., and Reinhorn, A.M. "A software for measuring disaster community resilience according to the PEOPLES methodology." *Proceedings of COMPDYN 2011—3rd International Conference in Computational Methods in Structural Dynamics and Earthquake Engineering*, Corfu`, Greece, May 26–28, 2011.

Bruneau, M., Chang, S., Eguchi, R., G. Lee, O'Rourke, T., Reinhorn, A. M., Shinozuka, M., Tierney, K., Wallace, W., and Winterfelt, D. v. 2003. A framework to Quantitatively Assess and Enhance the Seismic Resilience of Communities. *Earthquake Spectra*, Vol. **19**, 4, 733–752.

Bruneau, M., and Reinhorn, A. M. 2007. Exploring the Concept of Seismic Resilience for Acute Care Facilities. *Earthquake Spectra*, Vol. **23** 1, 41–62.

Chang, S., and Shinozuka, M. 2004. Measuring Improvements in the Disaster Resilience of Communities. *Earthquake Spectra*, Vol. **20**, 3, 739–755.

Cimellaro, G. P., Reinhorn, A. M., and Bruneau, M. 2005 Resilience of a health care facility. *Proceedings of Annual Meeting of The Asian Pacific Network of Centers for Earthquake Engineering Research (ANCER)*, Seogwipo KAL Hotel Jeju, Korea.

Cimellaro, G. P., Reinhorn, A. M., and Bruneau, M. 2006. Quantification of seismic resilience. *Proceedings of the 8th National Conference of Earthquake Engineering,* Vol. paper 1094, No., pp. April 18–22, 2006, San Francisco, California.

Cimellaro, G.P., Fumo, C., Reinhorn, A. M., and Bruneau, M. Quantification of Seismic Resilience of Health care facilities. 2009. *MCEER Technical Report-MCEER-09-0009.* Multidisciplinary Center for Earthquake Engineering Research, Buffalo, NY.

Cimellaro, G. P., Reinhorn, A. M., and Bruneau, M. 2010a. Seismic resilience of a hospital system. *Structure and Infrastructure Engineering*, Vol. **6**, 1–2, 127–144.

Cimellaro, G. P., Reinhorn, A. M., and Bruneau, M. 2010b. Framework for analytical quantification of disaster resilience. *Engineering Structures*, Vol. 32, No. 11, pp. 3639–3649.

Cimellaro, G. P., Christovasilis, I. P., Reinhorn, A. M., De-Stefano, A., and Kirova, T. 2010c. "L`Aquila Earthquake of April 6th, 2009 in Italy: Rebuilding a resilient city to multiple hazard." MCEER Technical Report – MCEER-10-0010, State University of New York at Buffalo (SUNY), Buffalo, New York.

Cimellaro, G. P., Reinhorn, A. M., and Bruneau, M. 2011. Performance-based metamodel for health care facilities. *Earthquake Engineering & Structural Dynamics,* Vol. **40**, pp. 1197–1217.

Cimellaro, G. P., and Kim, H. U. "The physical and economical dimension of Community Resilience "*Proceedings of the 2011 World Conference on Advances in Structural Engineering and Mechanics (ASEM11)*, September 18–23th, Sheraton Walker Hill Hotel, Seoul, South Korea.

Cimellaro, G. P. 2011. "From Performance-Based Design towards Resilience-Based Design." *Structure and Infrastructure Engineering*, in review.

Cornell, A., and Krawinkler, H. 2000. Progress and challenges in seismic performance assessment Peer News, Vol. April 2000, 3, No. 2.

FEMA. 1997. "FEMA 273 NEHRP Guidelines for Seismic Rehabilitation of Buildings." Federal Emergency Management Agency, Washington, D.C.

FEMA. 2000. "FEMA 356 Prestandard and Commentary for the seismic rehabilitation of buildings." Federal Emergency management Agency Washington DC.

FEMA. 2006. "FEMA 445: Next generation Performance-Based Seismic Design guidelines." Federal Emergency Management Agency, Washington D.C.

Garmezy, N. 1973. Competence and adaptation in adult schizophrenic patients and children at risk. In Dean, S. R. (Ed.), Schizophrenia: The first ten Dean Award Lectures (pp. 163–204). NY: MSS Information Corp.

Grossi, P. 2009. Property damage from the World Trade Center attack. Peace Economics, Peace Science, and Public Policy, Vol. **15**, No. 3.

Renschler, C., Frazier, A., Arendt, L., Cimellaro, G. P., Reinhorn, A. M., and Bruneau, M. 2010a Developing the "PEOPLES" resilience framework for defining and measuring disaster resilience at the community scale. *Proceedings of the 9th US National and 10th Canadian Conference on Earthquake Engineering (9USN/10CCEE),* Toronto, Canada, July 25–29, 2010.

Renschler, C., Frazier, A., Arendt, L., Cimellaro, G. P., Reinhorn, A. M., and Bruneau, M. 2010b. "Framework for Defining and Measuring Resilience at the Community Scale: The PEOPLES Resilience Framework." *MCEER Technical Report – MCEER-10-006*, pp. 91, University at Buffalo (SUNY), The State University of New York, Buffalo, New York.

Rose, A., Wei, D., and Wein, A. 2011. Economic impacts of the ShakeOut Scenario. *Earthquake Spectra*, Vol. 27, 539–557.

SPUR Hazard Mitigation Task Force. 2009. "The Dilemma of existing buildings: Private property, public risk." San Francisco Planning + Urban Research Association, February 1, 2009 (www.spur. org), San Francisco.

Tierney, K. 2009. "Disaster Responce: Research Findings and their implications for Resilience Measures." CARRI Research Report 6, Community & Regional Resilience Institute, Colorado Boulder.

Yang, T. Y., Moehle, J., Stojadinovic, B., and Kiureghian, A. D. 2009. Performance evaluation of structural systems: theory and implementation. *Journal of Structural Engineering, ASCE*, Vol. **(135)**, 10, 1146–1154.

Wein, A., and Rose, A. 2011. Economic resilience lessons from the ShakeOut Earthquake Scenario. Earthquake Spectra, Vol. **27**, 559–573.

Werner, E. E., and Smith, R. S. 1992. Overcoming the odds: High risk children from birth to adulthood, Cornell University Press, Ithaca, New York.

An auto-diagnosis tool to improve urban resilience: The RATP case study

M. Toubin
Egis, France
University Paris Est—EIVP, Paris, France
University Paris Diderot, Sorbonne Paris Cité, Paris, France

D. Serre
University Paris Est—EIVP, Paris, France

Y. Diab
University Paris Est, LEESU, Marne La Vallée, France

R. Laganier
University Paris Diderot, Sorbonne Paris Cité, Paris, France

ABSTRACT: Urban resilience is a new approach in urban development and risk management that focuses on disturbance absorption and functions recovery. Urban resilience is then based on urban services which support the cities functions and yet, display several critical issues in risk management. When considered as socio-technical systems in which physical network and governance are interacting, urban services are subject to continuity issues, mainly because of interdependencies. We propose here a collaborative method to help managers identify both technical and organizational dependencies so that they can share a common representation of the urban system and discuss its global resilience. The Parisian transportation system (RATP) illustrates the first part of the method: the auto-diagnosis tool to identify and prioritize dependencies. Then the rest of the method is shortly described and we outline the interest of the research in improving urban resilience.

1   INTRODUCTION

Every year, former records in damages due to natural hazards are exceeded, every year the insurers' reports warn governments about costs and economic consequences, and yet, every year, housings and activities are built in hazard-prone areas without adapted solutions. This paradox leads to the questioning of classical risk management and express the need for new approaches able to take into account long-term issues. Indeed, the conflict between short-term objectives: development or security and long-term objectives: sustainability and quality of life, is exacerbated in the urban context where most of society issues lie. Urban systems are threatened by many risks, both endogenous and exogenous (Godschalk 2003). Exogenous risks coming from the environment: the weather but also the relations with other urban systems or external networks, produce more and more damages in the urban context because of the concentration of exposed populations and goods, let alone decision-making centers or economic poles. Moreover, endogenous risks coming from the urban system itself are more and more responsible for service disruptions or even accidents as far as modern cities are heavily relying on urban networks and technologies. Then the urban system can be seen as a systemic model where five components are interacting: public infrastructures (authorities, hospital, schools and public services), activities, housing, population and technical systems or networks (lifelines, transportation systems). In this paper, we use equally the term urban service, technical network or system because we chose a systemic approach to assess urban resilience, so that technical networks are sub-systems of the urban system. These components perform urban functions that are essential to the global system functioning and the disruption of these functions are the consequences, or sometimes the cause, of disasters.

Then a new approach in risk management could consist in a better consideration of these functions, the components and their interaction within the urban system, in order to ensure the sustainability of cities (Comfort et al. 2010). This concept is called resilience and will not be defined in this paper as far as other works will discuss the concept in this book. Here, we first highlight the issues that must be tackled by this new approach in improving urban services resilience. The second part outlines the interests of collaborative approaches in doing so, and the last part gives an example of a tool to improve urban resilience.

2 A COMPLEX SYSTEM OF SYSTEMS REQUIRING NEW APPROACHES

2.1 *City's functions continuity*

Urban functions enable the city to develop and thrive; they are the basis of all other activities. Urban functions are characterizing a city and meet the basic needs of its population. Whether education, care, protection, supplies, information, culture, decision-making or planning, all urban functions need electricity, telecommunications, water or transport in order to operate normally. Then technical networks are essential to the city system, they are the backbone of its development (Bruneau et al. 2003) so that a resilient city must have resilient networks. Network managers have to comply with regulatory requirements concerning security and continuity of service. In doing so they have to thoroughly assess the functioning of their network, the critical points and to set up security plans to secure them. Then intrinsic vulnerabilities of networks are well known and quite well taken into account. But the main issue is to know and tackle interdependencies between networks.

Indeed, in the systemic model of urban systems, interactions between components have been highlighted; the same interrelations exist at a smaller scale between technical networks. Some of them need other resources in order to operate their service (Robert et al. 2009). And yet, these dependencies are not well known and scarcely taken into account. Functional dependencies are numerous inside the system of systems comprised of lifelines (Roe 2010). If some of them are apparent and well know (for instance the water pumps that need power), others are not that obvious and understood. Thus, the complexity of this system of systems (Eusgeld et al. 2011) and the growing scale of the connections (usually national, continental and even intercontinental) requires new approaches to tackle them and ensure the functioning of the global system.

2.2 *Interconnected networks and isolated managers*

If the internal functioning of each system is well known by the manager, the dependencies to external resources introduce uncertainty in its functioning and reliability. That's why in order to improve networks resilience, our work focuses on interdependencies (Lhomme et al. 2011). At the urban scale, technical networks are dependent from each other (straight arrows in Figure 1), but the many different managers in charge of one or several networks are not necessarily collaborating (curved arrows in the actors level).

On top of that, we can add the governance system that can be comprised of several institutions (especially in the French context), in charge of certain aspects, or delivering advice, means or requirements to the network managers. Usually urban networks are under the responsibility of the town services (or a public institution at the urban area scale, i.e. gathering several municipalities). But public services delegation has complicated the scheme and introduces new issues for network managers. Competitiveness, confidentiality and also security are now playing a part along with public interest and continuity of service that hitherto remains under the responsibility of public institutions.

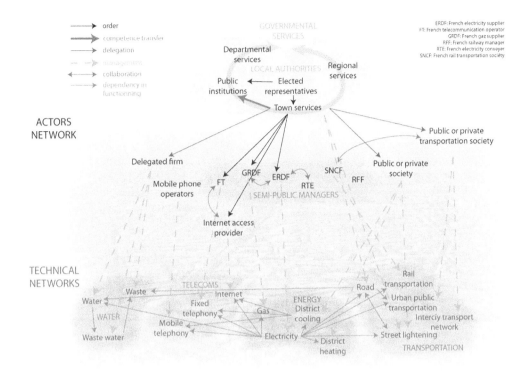

Figure 1.   Interconnected networks and isolated managers.

2.3   *Need for interdependencies identification*

In this context, the need for interdependencies identification is not easy to acknowledge for network managers. They are often reluctant to discuss their internal failures albeit they are prompt to blame other systems when they have not been able to operate their own because of a dependency. Indeed, for each system, many resources are needed so that they can provide other systems with their production. Electric power is the most central system as far as every other technical network needs power. But telecommunications have recently taken a place almost as important as electricity because many equipments are remotely controlled using telecommunication networks (Eusgeld et al. 2011). Drinkable water is indirectly needed by all other systems for the employees, but it can also be used in telecommunication centers for equipments cooling. The relative importance of those interdependencies needs to be assessed in order to prioritize the measures that have to be enforced. Analyses of the systems functioning, problem identification, solution definition and then implementation are time consuming, expensive and data may not be available to do so. In order to have an efficient policy in resilience improving, it is necessary to foster a collaborative approach concerning this issue.

3   RELEVANCE OF COLLABORATIVE APPROACHES

3.1   *Urban services as socio-technical objects*

If network managers implement solutions only at their network scale, they are likely to increase their own resilience, but decrease the global resilience. For instance, a classical strategy for electricity supply is to cut power in order to preserve the network, protect people and foster a quicker recovery. But if other systems (like urban heating, sewage system) need power to protect their own infrastructure, the balance between benefits and drawbacks is not obvious. Hence, the impacts of one solution on the resilience of the global system must be assessed so that it doesn't hamper another network or its capacity to face another disturbance. Beyond technical dependencies, organizational interactions must be taken into account as well. Indeed, strategies implemented by managers to ensure their continuity of service can hamper another service. On the contrary, managers could also mutualize means and knowledge when preparing for a crisis or during the recovery phase.

In this context, it seems necessary to consider a technical system as a system comprised of two components: the physical network and the service (managers), interacting with each other. Technical networks can then be regarded as a socio-technical object where technical functioning and human factors are influencing the global resilience. Thus, when trying to improve its resilience, it is necessary to act on both aspects and not only on technical interdependencies identification (Rinaldi et al. 2001; Ouyang & Dueñas-Osorio 2011). Collaborative approaches and in particular collaboration between experts and decision-makers could tackle those socio-technical issues.

3.2 *State of the art concerning collaborative approaches*

Collaborative approaches, generic term including all different scale of participation, have been developed with the rise of sustainability concepts and the need for more implication of citizens. Collaborative approaches can involve public and private interests, experts and decision-makers around problem definition or problem solving, assuming then that the problem is known and accepted by all stakeholders. The degree of involvement can range from simple communication, to coordination, cooperation and finally collaboration (Jankowski & Nyerges 2001). The objective is usually to let everyone express their interests, exchange information and then to agree on the definition of a problem or on the choice of solutions in order to solve conflicts (Ridder et al. 2005). They have first been tested and then theorized in resource management, particularly water, and are now widespread in all environment issues. And yet, both risks and networks management seem still dominated by experts and top-down approaches (see for instance the article by Arnaud *et al.* in this book). We admit that public participation may not be easy to take into account in managing technical networks (see for instance the analysis of the role of users in urban services in (Coutard & Pflieger 2002)), though many other experiments have demonstrated the interest of involving lay people in highly technical problems (Callon et al. 2001). Here is the difficult question of lay versus expert knowledge. Even if we decide to exclude public participation, the question of the knowledge provided or constructed to build the collaboration is not easy. Indeed, each manager has his own approach and science but we need a common representation of urban services that can be accepted and understood by all participants.

3.3 *Positioning and objectives*

Given the abovementioned issues and objectives in improving urban resilience, we argue that a collaborative tool to identify and discuss interdependencies between networks managers could be highly valuable. We will first involve managers, as experts in their system, and local authorities (decision-makers) which are in charge of the city's safety and continuity. The objective is to have them exchange information and perhaps find common solutions to common problems (Figure 2), which is the objective of collaboration as defined by (Jankowski & Nyerges 2001). We have already highlighted the lack of information concerning interdependencies identification and also the need for the construction of a common knowledge accepted by everyone. Then improving resilience would consist in identifying networks interdependencies with the managers in order to improve their knowledge and build a common representation of the urban system. Indeed, when facing such a complex object as the urban system, it is necessary to use cognitive, constructivist and participative approaches (Desthieux 2005).

This approach should help the managers and the decision-makers in understanding the interactions existing between the urban services so that they share a common knowledge of the system that can be the basis for collaboration. Problem identification can then be achieved during a collaborative workshop for instance, where managers can discuss the criticality

146

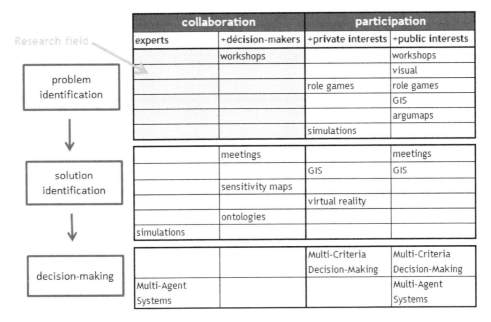

collaboration		participation	
experts	+décision-makers	+private interests	+public interests
	workshops		workshops
			visual
		role games	role games
			GIS
			argumaps
		simulations	
	meetings		meetings
		GIS	GIS
	sensitivity maps		
		virtual reality	
	ontologies		
simulations			
		Multi-Criteria Decision-Making	Multi-Criteria Decision-Making
Multi-Agent Systems			Multi-Agent Systems

Research field → problem identification → solution identification → decision-making

Figure 2. Positioning in the collaborative approaches and the different tools identified in the state of the art.

of their dependencies, express their needs and strategies (this part of the research is still ongoing and is just mentioned in 4.3).

4 METHODOLOGY AND FIRST RESULTS

4.1 *An auto-diagnosis tool to highlight dependencies*

Following the context, the issues and approaches identified in the former parts, we have developed a tool to highlight dependencies, raise awareness and feed a collaborative work between network managers. The tool is an auto-diagnosis method based on a spread sheet to be completed by a manager in order to help him identify the resources he needs to operate his system, and to identify the resources he provides to other systems. He is not asked to inform about the internal functioning but the resources he needs can be either internal or external. Considering that managers are reluctant (or sometimes not able) to share information concerning the internal functioning, we focused on dependencies to external systems, though we are convinced that the realization of the complete diagnosis would help the managers to identify internal failures. Several information concerning the resource are necessary: Who is the provider? Does the system dispose of autonomy concerning the resource? What are the consequences of a disruption? How long would it take to be repaired? They should in the end help the manager identifying the criticality of the resource. The same kind of information is asked concerning the provided resources: Who uses it? Is it necessary to them? How long would it take to be produced again? All in all, the manager has to imagine how his system would behave in case of disturbances and what would be the impact on its system but also on the systems that depend on it.

The fact that the diagnosis is to be filled by the manager fosters acceptance of the information so that the following collaborative work can take place on a legitimate basis. Besides, interesting results should emerge from the confrontation of the diagnoses. First, concerning the criticality estimated by the manager. It could appear that the user deems highly critical a resource he uses whereas the provider thought it was not indispensable for him. Secondly, the scenario in which the manager places himself will highlight imbalance between the different situations considered. The level of risk taken into account or the impacts imagined by the managers could be very different from one to the other and lead to inconsistencies in the preparation.

4.2 *The RATP case study*

The tool has first been experimented with the Parisian transportation society (RATP). The metro network is highly critical for the functioning of all the urban area, but it is also highly vulnerable to floods. Paris' urban area is threatened by the Seine's 100-year floods which latest event occurred in 1910, when the RATP networks already existed and were considerably damaged. The RATP planned for the crisis and now dispose of information on how the network behaved during the floods. The transportation network (especially the metros) is very old and highly vulnerable in case of floods. In order to preserve the infrastructure and the trains, which would be irreplaceable, the RATP have planned to protect the network. For each level of the river, they know which points are likely to be flooded. As a protective measure, they have planned to build walls around the station entrances (among others) in order to prevent the water from entering the underground. According to the RATP manager we met, we'll focus here on the metro system in the situation of moderated floods, i.e. where the network is not entirely flooded. In fact, 30% of the service should not be ensured because of the closed stations, mainly in the center part of Paris, near the Seine. Because of confidentiality matters, the auto-diagnosis is limited to the external resources and no information of quantities or location were given. For each resource, the manager identifies why it is needed, what autonomy the system has, the alternatives solutions that can be set up and then the criticality of the resource (Figure 3). For instance, he spontaneously added the resources needed for trains' maintenance. They come from several suppliers and the RATP disposes of a stock of spare parts (enabling a few weeks of autonomy) but it is necessary for the functioning of the system. The manager also identified several resources that are necessary for the employees to

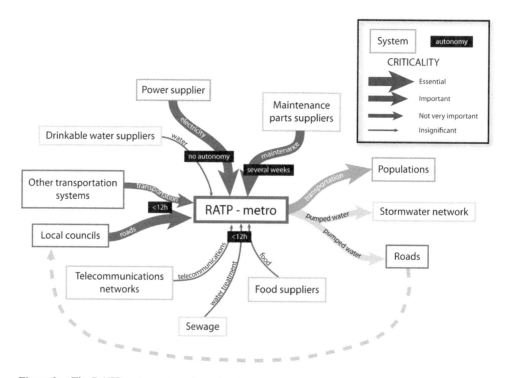

Figure 3.   The RATP-metro system dependencies.

be able to work: transportation and roads to come to work, food and water supply, sewage. Degraded conditions of these resources could probably be accepted during a few days in case of crisis but the impacts of a longer disruption should be assessed.

This graphical representation highlights the strong dependency on electricity (cf. 2.3) and the high vulnerability of the system to a possible disruption. As the manager notified, the RATP already know this critical dependency and they have set up a complete strategy to reduce its specific vulnerability. Another interesting criticality concerns the telecommunications network that is not as important as we might imagine. Indeed, the metro network uses its own telecommunication infrastructure to regulate the traffic or manage crises. We identified another interesting result of this analysis: an interdependency created by the system and symbolized by the dotted loop. Indeed, in order to protect its own infrastructure, the RATP plans to pump water and reject it in the streets or in the stormwater network. The amount of water rejected in the stormwater network is likely to be very small compared to the volume of an intense storm, but it is possible that in doing so, they bring water in areas that are not directly submerged (the location of which is known). The exact impact is studied by the RATP, it won't question the accessibility of these areas. Nevertheless, the possible retroaction loop was worth mentioning …

4.3 *Collaborative interdependencies learning*

With the first diagnosis, we begin to imagine the results that will emerge from the confrontation of all sector-based diagnoses. Inconsistencies or discrepancies might be highlighted but they should nurture a productive debate around the global system resilience. Based on the field experience of each manager, the interdependencies should be acknowledged by everyone and the temporal information will emphasize possible alternatives (autonomy, inertia, and resumption delays). Common problems are likely to lead to common solutions if the global interest is central. Besides, the multiplicity of points of view and interests is likely to raise questions of level of risk, unexpected (or even unimagined) disturbances and unlikely consequences. Indeed, if we resume the RATP case study, it appears that the level of risk considered (100-year floods) seems well managed. How about something goes wrong in their plan? How about the next Seine floods exceed this level? Obviously it is not possible to prepare for everything or at any level of risk, but the confrontation with other managers could be useful to question the risk management and step back. All in all, the objective is to discuss the city preparation to disturbances and find the better solutions not just at a network scale but at the city scale.

5 CONCLUSION

Urban resilience is enabled by urban functions resilience that relies on technical networks resilience. And yet, the global resilience is not the sum of sector-based resiliencies. Interconnections, dependencies, retroactions are complicating the urban system, so that the systemic functioning of all networks must be assessed. Collaborative approaches seem to be relevant to build a common knowledge and then foster an integrated management of urban services. The auto-diagnosis method proposed here aims at emphasizing the interdependencies between systems (both technical and organizational). Considering the issues specific to technical networks, we decided to limit the diagnosis to a global model, with little information about quantity, localization or internal functioning. With the first experiment on the RATP metro system, it seems that the objective can be achieved and will be able to fuel an interesting collaborative work between managers. We also remain convinced that more precise information about network functioning, dependencies and productions, such as geolocalization would enable managers to identify more precise solutions, at a smaller scale. And then, considering the population, activities and infrastructures which depend on the networks and located in the city, the local authority would have information to plan for a possible disturbance.

ACKNOWLEDGEMENTS

This research is part of the Project Resilis, led by Egis with the following consortium: EIVP, IOSIS, Sogreah, Cemagref, REEDS, LEESU, Fondaterra. (www.resilis.fr)

It is funded by the French National Research Agency (ANR Sustainable Cities 2009, reference ANR-09-VILL-0010-VILL).

REFERENCES

Bruneau, M. et al. 2003. A framework to quantitatively assess and enhance the seismic resilience of communities. *Earthquake Spectra* 19(4): 733–752.

Comfort, L.K., Boin, A. & Demchak, C.C. 2010. *Designing resilience : Preparing for extreme events.* Pittsburgh, USA: University of Pittsburgh Press.

Folke, C. et al. 2002. Resilience and sustainable development: building adaptive capacity in a world of transformations. In *World summit on sustainable development* on behalf of the Environmental advisory council to the Swedish Government: 34.

Godschalk, D.R. 2003. Urban Hazard Mitigation: Creating Resilient Cities. *Natural Hazards Review* 4(3): 136.

Holling, C.S. 1973. Resilience and stability of ecological systems. *Annual Review of ecology and systematics* 4: 23.

Lhomme, S., Serre, D., et al. 2011. A methodology to produce interdependent networks disturbance scenarios. In ASCE, ed. *International Conference on Vulnerability and Risk Analysis and Management, University of Maryland, Hyattsville, MD, USA*: 10.

Lhomme, S., Toubin, M., et al. 2011. From technical resilience toward urban services resilience. In E. Hollnagel, E. Rigaud, & D. Besnard (eds). *Fourth Resilience Engineering Symposium. Sophia Antipolis.* Presses des Mines: 172–178.

Ridder, D., Mostert, E. & Wolters, H.A. 2005. *Learning together to manage together—improving participation in water management.* Osnabrück, Allemagne: HarmoniCOP, Osnabrück University, Institute of environmental systems management: 99.

Robert, B. et al. 2009. *Organizational Resilience—Concepts and evaluation method.* Presses Internationales Polytechnique (ed.): 50.

Rose, A. 2011. Resilience and sustainability in the face of disasters. *Environmental Innovation and Societal Transitions* 1(1): 96–100.

Toubin, M. et al. 2011. Improve urban resilience by a shared diagnosis integrating technical evaluation and governance. *In EGU General Assembly 2011, Vienna, Austria.*

Resilience and Urban Risk Management – Serre, Barroca & Laganier (eds)
© 2013 Taylor & Francis Group, London, ISBN 978-0-415-62147-2

Flood resilience assessment of New Orleans neighborhood over time

M. Balsells
Université de Mons, Faculté d'Architecutre et d'Urbanisme, Mons, Belgium
Université Paris-Est, Ecole des Ingénieurs de la Ville de Paris, Paris, France

V. Becue
Université de Mons, Faculté d'Architecutre et d'Urbanisme, Mons, Belgium

B. Barroca
Université Paris Est, LEESU (Laboratoire Eau Environnement et Systèmes Urbains)—Département Génie Urbain, France

Y. Diab
Université Paris Est, LEESU (Laboratoire Eau Environnement et Systèmes Urbains)—Département Génie Urbain, France
Université Paris-Est, Ecole des Ingénieurs de la Ville de Paris, Paris, France

D. Serre
Université Paris-Est, Ecole des Ingénieurs de la Ville de Paris, Paris, France

ABSTRACT: New Orleans has coped with natural disasters since it is founding, but the most destructive was Hurricane Katrina in August 2005, both a natural and a man-made disaster. The resultant flooding demonstrates that the management of flood risk must incorporate new concepts like urban resilience in the city's planning process. The goal is to assess if resilience have been incorporated into the recovery of New Orleans post-Katrina, particularly of one city neighborhood: Oak Park. The assessment of Oak Park's urban resilience was made by comparing its pre and post Katrina conditions. A methodology was developed for modeling failure mechanisms by using two operational safety methods: a functional analysis and a failure mode and effect analysis. In New Orleans' post-Katrina recovery, a strong emphasis has been placed on improving the flood performance of buildings and protective systems but other components and vital functions of the neighborhood have not been fully considered.

1 INTRODUCTION

Floods are part of nature and they may affect all aspects of our lives. This is particularly the case for cities which are the most vulnerable because of the concentration in these areas of people, their possessions and their economics activities which are all subject to floods. Recent changes in urban systems and their environments caused by rapid urbanization and climate change increase both the probability and the impact of flooding. Moreover, there exist a high number of interdependencies between the individual components of the urban system that make cities more vulnerable to floods. During the last decade extremely damaging floods have occurred all over the world: New Orleans 2005, Central Europe 2009, Chine 2010, Thailand 2011, etc. These disasters highlight the necessity for a new and different approach to urban design, planning and building. Consequently, there is a need for all cities to adapt to climate and socio-economic changes by incorporating urban resilience strategies into overall city planning.

The goal of this paper is to assess the urban resilience of New Orleans during Katrina's flood and then analyze how its recovery or possible future projects could change its resilience.

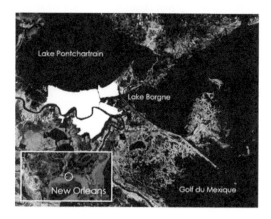

Figure 1.   Location of the New Orleans region.

In particular, we focus on one neighborhood in the city, Oak Park. We began by defining a neighborhood as a complex system and then develop a methodology for modeling its operation under flood conditions. Using the results of this analysis, we can then determine the tools needed to assess and discuss urban resilience of a neighborhood during a flood. Finally, by using these tools, we can apply them to an individual neighborhood in New Orleans which then allows us to assess the city's urban resilience under Katrina's flood conditions and to analyze how urban resilience concept has been incorporated in its recovery and how this concept can be introduced in future projects in the neighborhood and throughout the city.

2   THE CONTEXT OF NEW ORLEANS

New Orleans was founded in 1718 by the French Mississippi Company, under the direction of Jean-Baptiste Le Moyen de Bienville. The city is located in the Mississippi River Delta on the east and west banks of the lower Mississippi River and south of Lake Ponchartrain.

A really important feature of the city is its topographic elevation which is both father and son—producer and reflector, cause and effect—of the geology, pedology, hydrology and biology of New Orleans region (Campanella 2006). The average elevation of the city is currently between 1 and 0'5 meters below sea level, with some portions of the city as high as 6 meters above sea level (at the base of the riverfront) and other portions as low as 2 meters below sea level.

That New Orleans today is bowl-shaped and half below sea level is a result of soil subsidence induced by levee construction on the Mississippi River and drainage of the back swamp (Campanella 2006), and its topographic elevation is still diminishing.

Indeed, New Orleans is *"an inevitable city on an impossible site"* (Lewis 1973) which occupies a perilous place on a subsiding delta where flooding from the river, hurricanes and intense rain storms have been commonplace since the city's inception. It has survived 27 major floods over the past 290 years (Kates et al. 2006). By far, the most destructive and damaging was Hurricane Katrina in August 2005, which was both a natural and a man-made disaster.

When the levees protecting New Orleans failed, approximately 80% of the city was flooded. The extent of damages varied greatly from one part of the city to another. Some areas received 0,3 meters of water while others were submerged by more than 4,5 meters of water. The consequences of Katrina were staggering. Over 1600 fatalities occurred; there was over $20 billion in direct property losses with, 78 percent associated with residential losses. There was $10 billion in public property losses and indirect economic losses[*]. Beyond the fatalities, perhaps the most severe losses were to the social and cultural structure of the city.

---

* Greater New Orleans Community, http://www.gnocdc.org/

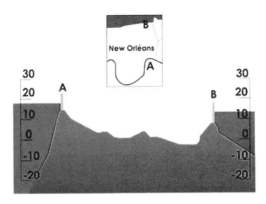

Figure 2. Topographic elevation of the New Orleans region.

Most of the neighborhoods that made up New Orleans were decimated, losing their population, social services, and social fabric.

Hurricane Katrina and the subsequent flooding demonstrate that man-made measures alone cannot sufficiently reduce risk for vulnerable areas such as New Orleans. And the risk is increasing significantly as a consequence of natural hazards such as hurricanes and other extreme weather events, which may now be stronger and more frequent due to climate change. Actually, climate change is expected to exacerbate the frequency and intensity of hydro meteorological disasters worldwide (IPPC 2007).

Since Hurricane Katrina, the U.S. Army Corps of Engineers has made significant improvements to the flood protection systems of the city. Currently, at 100-year HSDRRS (Hurricane & Storm Risk Reduction System) is being constructed which drastically reduces vulnerability to flooding for a majority of the region. This represent the best structural risk mitigation that the city has ever had, but it should be considered a baseline, not an endpoint because zero risk does not exist and it could be dangerous to believe otherwise (Serre 2011).

3 METHODOLOGY

3.1 *The concepts of resilience and urban resilience*

Derived from ecology, the concept of resilience is first defined as "the measure of the persistence of systems and of their ability to absorb change and disturbance and still maintain the same relationships between populations or state variables" (Holling 1973). This concept is used in many others disciplines (like psychology, economics, geography...) but for risk management this concept is relatively new, especially when it is applied to natural hazards.

Urban resilience can be defined as "the capacity of a city to face devastating events while reducing damage to a minimum" (Campanella 2006). This definition emphasizes the operational aspect of resilience that would tend to reduce the damage caused by a disturbance. However, there seems to be some confusion between the concept of persistence and resilience. After reviewing many definitions from different disciplines, a common definition of urban resilience emerges "the ability of a city to absorb disturbance and recover its functions after a disturbance". In other word resilience is the ability of a city to operate in a degraded mode and to recover its functions while some urban components are disrupted (Lhomme et al. 2010).

3.2 *Systemic approach of a neighborhood*

A city is an urban system spatially composed of individual neighborhoods. A neighborhood is a specific spatial entity providing several functions: housing, social services (education, health care, recreation...), city services (police, fire, sanitation...), commercial activities, etc. which together lead to specific activities and flows. Urban functions include all of the city's activities.

Individual neighborhoods can and do offer a specific number of these functions and in varying degrees or mixes. Actually, the urban functions of a neighborhood depend on its location and geography within the city, its unique physical and social history, its population profile (racial composition, age and economic mix density…), its land use configuration, etc.

The concepts of functional and social mix are important goals for a sustainable urban development. Since several urban functions are represented in a heterogeneous neighborhood, specific urban dynamics can be identified within it. Discussing and modeling an urban system at another level other than the city level is also possible: that is, at the neighborhood level. Indeed, in the same way as a city, a neighborhood can be considered as a complex open system which is characterized by exchange processes within its environment and is continuously changing and developing.

In order to asses a neighborhood's urban resilience under flooding conditions, we consider a neighborhood as a complex system and we use previous work done on the analysis of the city as the basis for our urban system modeling at this neighborhood level. By using this approach, it is possible to identify a system whose main urban components include: networks (water, energy, transportation, and telecommunication), public/institutional facilities, protective systems, housing and commercial.

These eight main categories of components when taken together create the physical environment of the neighborhood. This systemic modeling (Fig. 3) provides the working basis for the analysis we propose as follows.

3.3 *Operational safety method analysis for complex systems modeling*

To improve our understanding of how a neighborhood works and asses its performance, we begin by developing a methodology for modeling failure mechanisms (failure mechanism is the relationship between causes and effects leading to the failure of the neighborhood) using the following operational safety methods: functional analysis (FA) and failure mode and effects analysis (FMEA).

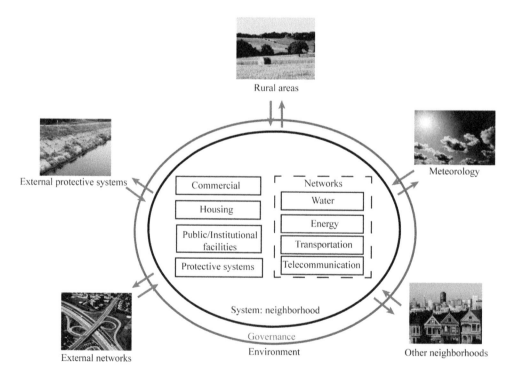

Figure 3. Systemic approach of a neighborhood.

154

The operational safety methods were primarily developed to study industrial systems in complex operations (important number of components, multiple and looped failures...) for which it is very difficult or impossible to produce operational modeling by conventional physical approaches (Serre et al. 2008).

In the case of an urban neighborhood, we are faced with a complex system such it was described above. Consequently, the operational safety methods seem well suited for modeling the operation of an urban neighborhood.

3.3.1 *Functional analysis*

The first method used in an operational safety exercise is a functional analysis: a systemic method to understand and describe how a system works. It provides us with a complete analysis of the neighborhood's components and their interactions with each other and the outside world and consequently the functions they provide. The components' interactions with each other and with the outside world are mapped as functional block diagrams (Fig. 4).

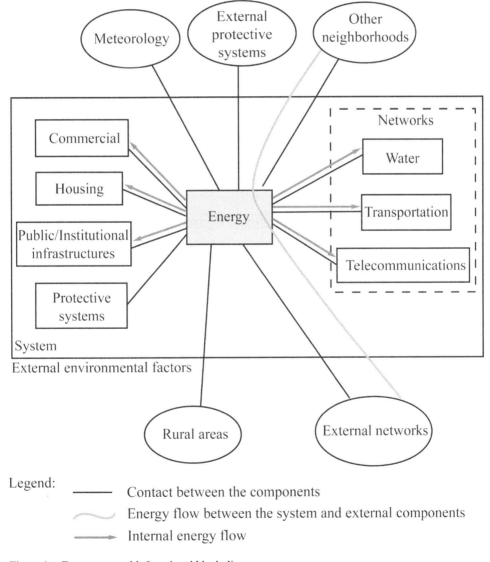

Figure 4. Energy network's functional block diagram.

155

The functional block diagram shown in the Figure 4 indicates contact and energy flow relationships between the energy network and other components of the neighborhood (commercial, housing, etc.) and external environment factors (rural areas, external networks, etc.). We note that the energy network is particularly important because the normal operation of all other components depends on it. By means of these results, we then can distinguish and highlight the design functions of the energy network.

3.3.2 *Failure mode and effects analysis*

The second method to be used is the FMEA, which is an inductive method for analyzing potential system failures. It focuses systematically on each component in the system to determine its failure modes (a failure mode is the nonperformance of a function under system conditions: absence, loss, deterioration, or untimely operation of a function and the effects and consequences thereof), and their causes and effects.

Applying FMEA for the purpose of this study, we can determine the more important failure modes of the neighborhood's components, their possible causes and their possible effects under flooding conditions.

The Table 1 shows the FMEA results for the energy network. In particular, the table shows the component and its schedule number, its design functions, its failure modes and possible causes and effects of its failure during a flood. For example, the function *"Provide housing, commercial and public/institutional infrastructures with energy"* (column 3, Essential function) can cease to function due to a failure or deterioration of its primary function (column 4, Failure mode) caused by the corrosive effects of flood waters (column 5, Possible cause of the failure). A possible effect of this failure (column 6, Possible effect of the failure) could be the corrosion of network's elements (technical failure) or that housing, commercial and public/institutional infrastructures aren't served by the network (essential failure).

Consequently, the FMEA provides a good understanding of multiple and complex disorders that can result from the flooding of a neighborhood.

3.3.3 *FA and FMEA results*

By using the functional analysis (FA) we can illustrate both high dependencies and interdependencies between some of the neighborhood's components. The failure mode and effects analysis (FMEA) shows that these dependences become real issues under flooding conditions: there is a domino effect between neighborhood components due to their indirect

Table 1. Energy network's FMEA.

Nº	Component	Component's design functions		Failure mode		Possible cause of the failure	Possible effect of the failure	
		Essential	Technical	Technical functions	Essential functions		Technical failure	Essential failure
2	Energy network	*Provide flows* - Provide housing, commercial and public/instituti onal facilities with energy (electricity, natural gas) - Provide the other networks which depend on energy (electricity, natural gas...)	*Withstand the mechanical stress (multi hazards)* - Withstand (pipes) the technologic accidents - Ensure the service's continuity - Be sustainable over the time *Withstand the external factors (multi hazard)* - Withstand the different meteorology conditions (storms, snow, wind...) - Be waterproof - Withstand the energy failures (electricity) - Ensure the service's continuity - Be sustainable over the time	Failure or deterioratio n of the function "Withstand the mechanical stress (multi hazards)" Failure or deterioratio n of the function "Withstand the external factors (multi hazards)"	Failure or deteriorati on of the function "Provide flows"	- Corrosive effects of flood waters - Soil subsidence - Uprooting of trees - Infiltration of the flood waters in the network	- Corrosion of network's elements (pipes, pumps, valves...) - Change in network's elements (pipes) restraining conditions - Paralysis of network's elements (valves, pumps...) - Deformation , cracking or destruction of the network's elements	- Housing, commercial and public/instit utional facilities aren't served by the network - Damaged to networks which depend on energy (electricity)

156

vulnerability and the huge difficulties in recovery post-flood as a consequence of specific recovery dependencies.

3.4 *Urban resilience assessment tools*

Using these analytic methods and considering the definition of urban resilience (Lhomme et al. 2010) as quoted above, we conclude that in order to assess the urban resilience of a neighborhood it is necessary to study the following: the performance, indirect vulnerabilities and recovery dependencies of the individual neighborhood components.

The performance of individual neighborhood components is evaluated by using performance indicators. Indicators describe the effects of a specific phenomenon; in our particular study, they will provide specific information about the effects of flooding: i.e. how the failures of the functions are observed on the neighborhood components.

We can then proceed to define performance indicators in order to assess how each component of the neighborhood performs under flood conditions. These are defined by the possible effects of the failure of neighborhood components which were determined in the FMEA. This, in turn, results in 23 different performance indicators.

As an example, Table 2 shows the performance indicators which have been defined for the energy network. When the indicator is defined, it should be used. There are consequently no good or bad indicators, but they can be more or less adapted to characterize a specific phenomenon. During its existence, an indicator it may be modified, challenged and even abandoned. The use of an indicator depends essentially on the willingness of specific actors to use it (Bonnefous et al. 2001).

Based on FMEA's results we also propose possible indirect vulnerabilities and recovery dependencies of the neighborhood components, allowing them to be assessed under flood conditions. It is really important to identify and asses the indirect vulnerabilities and recovery dependencies of the neighborhood components during a flood because by reducing their impact, the neighborhood can operate in a degraded mode and its ability to recover its functions after a disturbance will increase.

Finally, we assess the urban resilience of the neighbourhood based on the following:

1. By using the performance indicators assessment, we can define how their performance can be detected, identify the type of assessment (qualitative or quantitative) and assign a performance score to each one by means of a scale of scores (Tab. 3), as well as any hypotheses and/or qualifications that are proposed.

    Some of the performance indicator qualifications are based on expert knowledge. It is also important to realize that any performance indicator which provides information about time period cannot be qualified due to the number of variables affect its value.

Table 2. Energy network's performance indicators.

N°	Indicators
1	Status of the network's elements
2	% of network damaged
3	Time required for network's recovery
4	% of users not served by the network
5	Time to return the network's service to users
…	…

Table 3. Scale of scores.

Score	Qualification	Color code
1	good	
2	fair	
3	poor	
4	very poor	

2. Next, by using the indirect vulnerabilities assessment, we can propose an evaluation matrix with three different values: low, medium and high.
3. Finally, by using the recovery dependency assessment, we can also prepare an evaluation matrix with three different levels of recovery dependency: low, medium and high.

As a result, we now have the tools to discuss a neighbourhood's urban resilience.

4 RESULTS AND DISCUSSION

4.1 *Neighborhood model: Oak Park*

The study area is an urban neighborhood located within Gentilly, which is a larger area in New Orleans bounded by St. Bernard Ave, Paris Ave, Robert E Lee Blvd, Mirabeau Ave. Oak Park is located near the south shore of Lake Pontchartrain and the University of New Orleans.

Oak Park is an area of low-lying ground composed, mostly of single-family residential built as slab-on-grade houses. By applying the systemic modeling defined above to Oak Park, its different urban components can be identified: 899 residential properties, 1 commercial center, 3 churches, 3 schools, a pair of I-Wall panels on each side of the London Avenue Canal as well as water, energy, transportation and telecommunication networks. Although the London Avenue Canal is not in Oak Park, its I-wall panels served as the protective system for the neighborhood and directly impacted its safety and security.

Pre-Katrina, Oak Park residents considered their community stable and attractive. The general condition of its networks (water, energy...) was good and the services provided for the residents were adequate. However, some problems existed: streets were in very poor condition, particularly internal streets which suffered severe subsidence pre-flood. The general condition of the buildings in the neighborhood was good although they were not elevated.

Katrina flooded Oak Park on August 29, 2005 when portions of the western I-wall panels of the London Avenue Canal were breached, resulting in 8 feet of water over 100% of the neighborhood. Using the tools mentioned above (performance indicators, indirect vulnerabilities and recovery dependences), we can now discuss the urban resilience of Oak Park during Katrina's flood and we can also analyze how its recovery or future projects could improve the neighborhood's urban resilience.

4.2 *Resilience of Oak Park under Katrina flooding conditions*

First, what was the urban resilience of Oak Park during the flooding caused by Hurricane Katrina and resultant flood wall failure? By all performance indicators it ranks very poor: 100% of the networks were damaged, 99% of buildings were damaged by 50% or more, 70% loss of the tree canopy, etc.

As one example, the energy network was completely damaged (100%). Several elements of the network were in a really poor status: electricity lines snapped and poles cracked; there was gas pipes corroded; the electricity converters were destroyed; 100% of the other Oak Park's components were not served by the network. Consequently, the score chosen for all indicators is very poor, as is shown in the Table 4.

There were also high indirect vulnerabilities and recovery dependencies in the neighborhood's components. Figure 5 shows how the online components were vulnerable to the failure of the components in columns during the flood due to Katrina. Among the results we can notice a high vulnerability of all components of Oak Park to the failure of protective systems and a high vulnerability of housing, commercial and public/institutional facilities to the failure of networks. Indeed, when the I-walls of the London Avenue Canal pushed over, the flood waters overflowed into Oak Park causing great damage to all components. As a consequence of the networks' failure housing, commercial and public/institutional facilities didn't have a "normal" operation.

Table 4. Performance indicators' qualification under Katrina flooding conditions.

N°	Component	Indicator	Score under Katrina flooding conditions
2	Energy network	Status of the network's elements	4
		% of network damaged	4
		Time required for network recovery	
		% of users not served by the network	4
		Time to return the network's service to users	

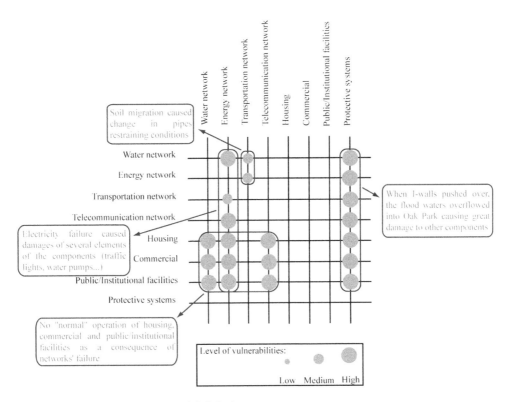

Figure 5. Indirect vulnerabilities of Oak Park components.

We should also highlight the high vulnerability of all components to the failure of the energy network; particularly the electricity failure which caused damage to traffic lights, water pumps... Indeed, electricity had and has a very important role in the overall neighborhood operation.

Figure 6 shows how the online components were dependents on the components in column to recover after the flood due to Katrina. Among the results we can notice a high dependence of all components needing to be recovered on the energy, telecommunication networks and

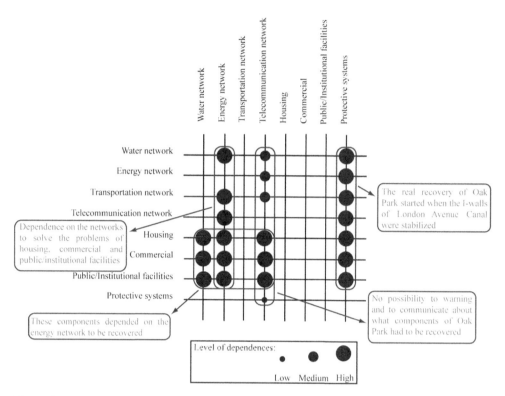

Figure 6. Recovery dependencies of Oak Park components.

on the protective systems. Actually, the real recovery of Oak Park started when the I-walls of the London Avenue Canal were stabilized. At the time, there was no way to communicate about what components of the neighborhood had been restored and simultaneously all components except protective systems couldn't be restored until the energy network, particularly electricity, was reestablished.

We should also highlight the high dependence on the water, energy and telecommunication networks to solve the problems of housing, commercial and public/institutional facilities. Indeed, they couldn't have a "normal" operation until these networks were recovered.

By using performance indicators, indirect vulnerabilities and the recovery dependency assessment, we can conclude that the neighborhood was not resilient during Katrina's flooding. All components in Oak Park performed poorly. As a consequence of the high indirect vulnerabilities of the components, Oak Park did not have the ability to operate due to the disruption of some components even, in a degraded mode. Oak Park did not have the ability to recover its normal functions after the disturbance due to the high recovery dependences of all components.

Indeed, urban resilience had not been incorporated into the urban development in Oak Park. Rather, the neighborhood relied on a system of levees, flood walls, and pumps to provide primary flood protection. Everyone in Oak Park assumed they were safe in their neighborhood and in the city based upon the flood protection systems.

4.3   *Oak Park resilience of post-Katrina recovery*

In this section, we describe the neighborhood's recovery to date, highlighting the main differences between pre-Katrina and post-Katrina conditions. We then analyze how these differences could improve the urban resilience of Oak Park.

Internal streets of Oak Park (roadway, sidewalks...) have been repaired 0%, basically they remain untouched

118 residential properties (13%) are still uninhabitable

Some churches remain as they were just after the flood

Oak Park's commercial center is largely vacant

School under development

Figure 7.   Some restorations which need to be made in Oak Park.

Figure 8.   Buildings elevated in Oak Park.

Table 5.   Commercial's performance indicators whose score could be improved.

N°	Component	Indicator	Score under Katrina flooding conditions
6	Commercial	% of buildings totally damaged	4
		% of buildings minimally damaged	4
		Time required for buildings recovery	
		% of buildings rehabbed	4
		% of commercial properties operational	4
		Time required to return to "normal" commercial operations	

Figure 9.   Pump station and storm surge gate at the London Avenue Canal.

The recovery of most of the neighborhood's components has been made to pre-Katrina conditions. Today there are two significant differences: the buildings are now elevated and the protective systems have been improved.

4.3.1   *Buildings*
Post-Katrina, new buildings codes have been adopted and a significant number of buildings are now elevated in Oak Park. These buildings are more resistant and have been adapted to flooding.

4.3.2   *Protective systems*
Two new elements have been incorporated into the protective systems of Oak Park: a pump station and a storm surge gate have been constructed at the outfall of London Avenue Canal. Moreover, the flood walls of London Avenue Canal have been repaired. These protective systems dramatically reduce the vulnerability of flooding for Oak Park; they represent the best structural mitigation for neighborhood.

With the description of Oak Park's recovery complete, we now proceed to analyze how it has affected the urban resilience of the neighborhood and study possible changes in the three different tools used in this case study.

First, the performance of some components of Oak Park could be improved. New elevated buildings could enhance the performance of housing, commercial.

The two new elements incorporated into the protective systems could further enhance the neighborhood's performance, by improving the qualification of their performance indicators.

Next, as is highlighted in Figure 10, some indirect vulnerabilities could be reduced as a result of the new buildings. Elevated housing, commercial and public/institutional facilities could be less vulnerable to the failure of protective systems. These elevated buildings are more adapted to flooding; if protective systems fail and flood waters overflow into Oak Park the damage caused to buildings, commercial and public/institutional facilities be reduced.

It should be noted that the new pump station and surge gate that have been incorporated into the overall flood protective system do not affect indirect vulnerabilities of the neighborhood's components.

Finally, recovery dependences could be also affected by the new buildings in Oak Park. Housing, commercial and public/institutional facilities could be less dependent on the protective systems in order to recover (Fig. 11). Even if there is water in the neighborhood because of a failure in the flood protective systems, these components could recover their functions more easily.

As a result, we have determined that Oak Park has improved its urban resilience. However, vital functions of the neighborhood have not yet been considered in order to more improve indirect vulnerabilities and recovery dependencies of its components. The reconstruction of Oak Park has been very fragmented and has lacked an overall vision of the neighborhood. It is also important to highlight the strong relationship between the solutions proposed to improve the urban resilience of Oak Park and its topographic elevation.

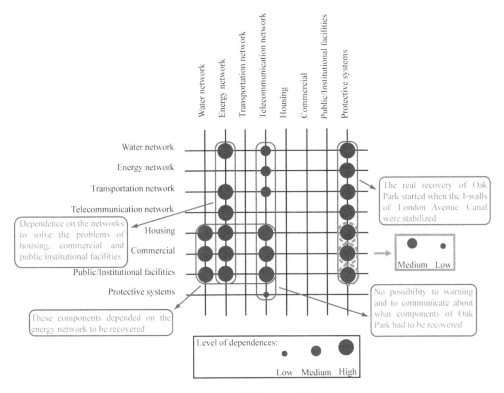

Figure 10. Possible changes of indirect vulnerabilities post-Katrina recovery.

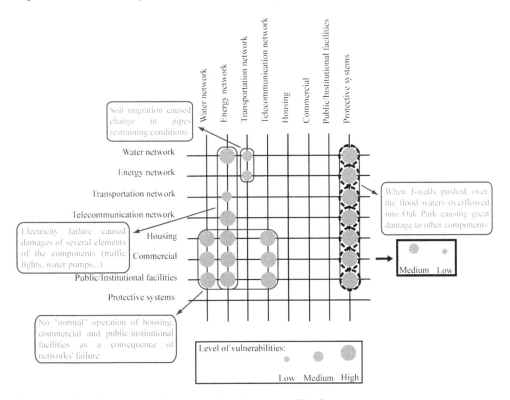

Figure 11. Possible changes of recovery dependences post-Katrina recovery.

4.4 Possible future Oak Park's resilience

A new concept has emerged post-Katrina that could significantly impact future projects for Oak Park: "living with water". This concept looks at new opportunities to create additional water storage capacity in the neighborhood while lowering the risk of storm water flooding during hurricanes. One example, Figure 12 shows a future project proposed for the neighborhood.

These possible future projects will not affect the performance of the Oak Park's components and the recovery dependences. However, they could reduce some indirect vulnerabilities. In particular, all components could be less vulnerable to the failure of the protective systems (Fig. 13).

The impact of the flood waters on the components could be lower because flood waters will find the place in the neighborhood. Moreover, they will compliment and provide long term solutions to water management by reintroducing water as a positive element in the neighborhood.

Figure 12. London Ave Canal existing and proposed section (Waggonner D. & Ball Architects).

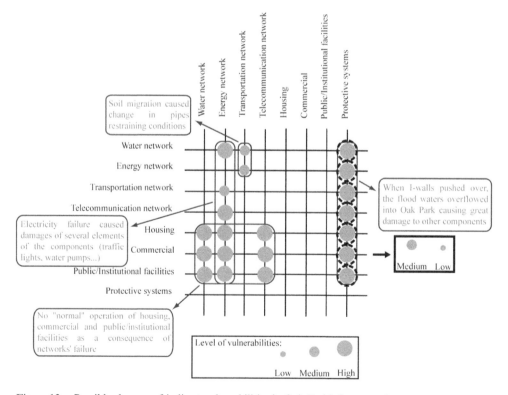

Figure 13. Possible changes of indirect vulnerabilities in Oak Park's future projects.

To conclude Oak Park could be more resilient in the future if these projects are carried out. However, what can be done to further improve the performance, indirect vulnerabilities and recovery dependencies of the neighborhood components have yet to be considered. Indeed, the urban components of the neighborhood are still considered separately.

5 CONCLUSIONS

A neighborhood is a set of urban components interacting in space and time. In order to study the urban resilience of a neighborhood a systemic approach has been used. The first stage used operational safety methods: functional analysis and failure mode and effects analysis. These methods made it possible to understand the functions of neighborhood's components and, identify their failures modes and determine causes and effects under flood conditions. In terms of results, based on these analyses, we have determined that in order to assess urban resilience of a neighborhood it is necessary to study performance indicators, indirect vulnerabilities and recovery dependences of its components. The next stage of our research consisted of defining performance indicators, indirect vulnerabilities and recovery dependences of neighborhood's components assessed under flood conditions and determine the neighborhood's urban resilience.

Finally, when applied to a particular neighborhood in New Orleans, Oak Park, we have seen that it was not resilient during Katrina's flooding and in its post-Katrina recovery resilience is only being applied to individual buildings and protective systems. There is not an overall vision of the neighborhood. Future projects for Oak Park could introduce water as a positive element in the neighborhood while reducing some indirect vulnerabilities of its components under flooding conditions. However, many practices have yet to be proposed to incorporate resilience into all of its networks while allowing all of its components to continue functioning in the face of rising water.

To conclude, a perspective of our research is improving the systemic modeling of urban system at the neighborhood level while introducing the flood system into the methodology. This model will allow us to better analyze the urban resilience of a neighborhood and to define specific features of flood resilience into urban design, planning and building. An interdisciplinary approach, particularly in the preliminary stages of urban design, will be required to create a highly urban resilient neighborhood.

REFERENCES

Beacon of Hope, Resource Center 2011. Gentilly open house report. http://www.beaconofhopenola.org/
Bonnefous, C. & Courtois A. 2001. Indicateurs de performance. Paris: Editions HERMES Science Publication, 285 p.
Campanella T. J. 2006. Urban Resilience and the Recovery of New Orleans. *American Planning Association. Journal of the American Planning Association, ProQuest Direct Complete*, pp. 141–146.
Campanella R. 2006. Geographies of New Orleans, Urban fabrics before de Storm. Center of Louisiana Studies, pp. 45–49.
CHART: Center for Hazards Assessment, Response and Technology. http://www.chart.uno.edu
Colten C. E. 2008. Community resilience: lessons from New Orleans and Hurricane Katrina. CARRI Research Report 3, 1 p.
FOLKE C. 2006. Resilience: The emergence of a perspective for social–ecological systems analyses. *Global Environmental Change*, vol.16, n°3, pp. 253–267.
Flood risk in New Orleans, implications for management and insurability 2006. Risk Management Solutions, Inc, 2 p.
Gunderson L.H. & Holling C.S. 2002. Panarchy: Understanding Transformations in Human and Natural Systems. Island Press, 508 p.
Holling C.S. 1973. Resilience and stability of ecological systems. Annu. Rev. Ecol. Systemat. 4 p. 1–23.
IPCC (Intergovernmental Panel on Climate Change) 2007. *Climate change 2007: The IPCC fourth assessment Report, Summary for Policymakers*, www.grida.no/climate.

Lewis E. Link 2009. Water Ressources Policy and Practice Issues Exposed by Katrina. *Journal of Contemporary water research & education issue* 141, pp. 9–14.

Lhomme S., Serre D., Diab Y. & Laganier R. 2010. Les réseaux techniques face aux inondations ou comment définir des indicateurs de performance de ces réseaux pour évaluer la résilience urbaine. *Bulletin de l'association des géographes français.*

Lhomme S., Serre D., Diab Y. & Laganier R. 2011. L'étude des effets dominos de systèmes fortement interdépendants à l'aide de méthods de Sûreté de Fonctionnement. *XXIXe Rencontres Universitaires de Génie Civil.*

Pasche, E. & Geisler, T. 2005. New Strategies of Damage Reduction in Urban Areas Prone to Flood, in Urban Flood Management, Szöllösi-Nagy, A. & Zevenbergen, C. (Eds), A.A. Balkema, Leiden, Netherland.

Parsons K. 2010. Our big bike plans and neighborhood recovery. *APA Mobile Workshop #W20.*

Sewerage & Water Board of New Orleans 2006. *Report on Current and Future Capital Needs*, pp. 15–30.

Serre D., Peyras L., Tourment R. & Diab Y. 2008. Levee performance assessment:development of a GIS tool to support planning maintenance actions. *Journal of Infrastructures Systems, ASCE*, Vol. 14, Issue 3, pp. 201–213.

Waggonner D. 2009. New Orleans/Netherlands: Common Challenges in Urbanized Deltas. *An anthology of writings about the proceedings of the Dutch Dialogues Conferences.*

Resilience and Urban Risk Management – Serre, Barroca & Laganier (eds)
© 2013 Taylor & Francis Group, London, ISBN 978-0-415-62147-2

Resilience: A pragmatic approach to the concept through the creation of the Ile-de-France (IdF) regional master plan

Mireille Ferri

Former Vice-Chairwoman of the IdF Regional Council, in charge of territorial development
and the regional master plan, Vice chairwoman of the Ile-de-France Development and Town Planning
Institute (IAU)

The concept of resilience applied to the urban world has suddenly materialized recently, containing a host of different meanings, as it stems from various different disciplines: From physics at the very beginning of the 20th century, to ecology and neurosciences, where it has been democratized by Boris Cyrulnik. It undoubtedly corresponds to the need for new approaches to be found and for it to be illustrated by new terms which represent notions of security for societies in the face of fears about a future that is more uncertain than ever.

As an echo to current research and widely presented in this document, I will come back later to the adventure of creating a new master plan for the Ile-de-France region, a new democratic and technical process for the Regional Council, carried out between 2004 and 2010. At the time, the "resilience" concept was not a part of our vocabulary. Even so, two key concepts were introduced at the very beginning of the exercise. The first, "robustness", was to throw light on the meaning of an approach totally different to previous regional strategic planning exercises in the Ile-de-France. The second did not have the same fate. "Flexibility", which should have been the complement to robustness, foundered very rapidly. Robustness + flexibility ... even so, it is by an attempt to create this equation that the notion of resilience can be mobilized today. Resilience makes us aware of the changes we are undergoing, which are caused by the energy question, leading to environmental consequences of unparalleled severity, by global upheaval of social and economic systems, by more or less chaotic and brutal reorganization of human migratory flows, by exchanges of physical assets and resources and by spatial, social democratic modes of organization ... capable of resisting all "types" of impact, of preserving sites and strategic functions and of rebuilding (robustness). To be able to adapt ceaselessly by finding the resources and strength for the rebuilding operation (flexibility).

When the master plan was introduced, three basic challenges were announced: The challenge of building "social robustness", which must consolidate cohesion between populations and territories. The challenge of building "environmental robustness", which will invent new processes for making products and services, and which will firmly reintroduce short cycles and local cycles into our territorial organizations. The challenge of building "economic robustness", which enables the way in which exchanges are organized and in which values are produced to be thought up differently, more especially because the third challenge takes account of the need to raise the first two challenges!

In my opinion, this approach is directly linked to introducing this still "original" notion of resilience into the orientations, objectives or implementation tools for the master plan.

1 PROPOSING ACTION WITHIN A SPATIAL FRAMEWORK

Intervening in Ile-de-France territories from the point of view of global protection for society that is inevitably confronted by today's changes means taking a new look at the way in which this regional system is organized in itself.

The SDRIF sets the organization of a "system" as the guiding principle and then develops all the implications involved in this view. The notion of a system appears as soon as the challenges are revealed and it is clearly linked with robustness: "Reinforcing robustness vis-à-vis climatic fluctuations means reinforcing the Ile-de-France system's general robustness". On this basis, development and town planning need to be thought out differently in order to preserve vital resources and use them rationally—in whose front line we find space and land—to anticipate risks and protect ourselves from them and to build sustainably.[1]

> "Reinforcing robustness vis-à-vis climatic fluctuations means reinforcing the Ile-de-France system's general robustness". On this basis, development and town planning need to be thought out differently in order to preserve vital resources and use them rationally, in whose front line we find space and land. (...)"

The program is vast and most certainly more than a little out of proportion to the SDRIF's means of creation and implementation Nevertheless, including this type of chapter in the document's introduction reflects the evolution that has been observed with the progression of debates which have mobilized experts, politicians and citizens in many Ile-de-France communities for several months. On the basis of this emerging notion, which had not yet been christened "resilience", we were able to build the beginnings of a common culture, still fragile because it is still only shared on the scale of a too small number of players, but whose progress can be identified in a number of documents on town planning or in places of reflection. We more especially saw the International Consultation on Greater post-Kyoto Paris take over a number of leads proposed in the SDRIF's development work. At present, the tools that appropriate this work are few and far between. New statutory frameworks and, above all, new partnership and financial arrangements must appear for real large-scale initiatives to take place.

2 INTEGRATING RISK BY INNOVATING URBAN RESPONSES AND ASSESSMENTS OF GLOBAL ROBUSTNESS

2.1 A local approach to the notion of risk

Even if, when reading the SDRIF, the paragraphs devoted to natural risks are globally "traditional" and their headings represent a continuity of existing actions, especially concerning links with reducing vulnerability situations, there is still a significant innovation.

Running counter to the notion of limiting risks, we have tried to introduce possibilities of testing new forms of construction, of network design, of ways of testing materials, etc. on a number of experimental zones. The purpose of this approach is to learn how to face up to known risks by incorporating their size and/or their frequency. From this point of view, flooding seemed to be the most appropriate hazard. Therefore, the SDRIF indicates (Page 111):

> "(...) the hypotheses on climate change made at present—and the phenomena observed over the last few years in Europe (Prague) and in France (Somme)—show that even more exceptional floods are likely. (...) Future developments must be compatible with flood risks. The best solution would be to favour the development of urban responses, which, in the context of measures contained in FRPP2, would enable town planning to be oriented towards less risk-generating systems and a reduction to be made in the vulnerability levels of equipment and housing. Innovating building operations could be developed in these zones, acting as full-scale tests."

Unfortunately this measure received a very negative reaction from the Ministry of Ecology, which considered that, in this case, the SDRIF did not comply with principles of precaution.

---

1. SDRIF – Page 38.
2. Flood Risk Prevention Plans.

The SDRIF's environmental assessment is an innovation: In Europe there are very few assessments on the scale of large urban metropolis that refer to planning documents and concern environmental issues.

Using two facts objectively recognized at the beginning of the process as a basis, the increase in globalization and the climate change, the assessment approach aims at measuring the impacts of not only a set of projects, but also of global strategy. As a result, this break in the ways territorial planning is tackled has been endowed with a new tool that seems to me to be inseparable from a forward-looking reflection on the capacities of a system to face up to identified hazards and to anticipate methods of recovery and associated tools needed afterwards.

In this way, environmental assessment refers directly to this will to reinforce the global system subject to uncertainty and to the need to anticipate, in a paragraph entitled "*Robustness: In front of the increasing uncertainties with which the whole territory is faced, the future must be thought about, not in terms of sector-based performances placed side-by-side but in terms of global robustness. It is a question of integrating uncertainties and changes, of reducing the territory's vulnerability, of applying principles of precaution and of reconciling economic performance with social cohesion and environmental protection.*"[3]

3   DEMOCRATIC EXPERIMENTING AS AN INTEGRATOR OF URBAN EVOLUTION

The master plan did not lay down the need for "robustness" as an invariant: It connects it to the imminence of change. In the face of a radical change in the physical, social or economic environment, it is essential to act on a large scale and within a short time, or, in any case, much shorter than the time normally associated modes of evolution in forms and organizations. Therefore, massive mobilization of different energies is needed: To make this type of effect, prodigious "social forces" are needed, whence the importance of re-democratizing our societies, beginning at the most local points. This schematic reasoning encourages development of new forms of transmission of information, renewal of decision-taking processes and reinforcement of both monitoring and assessment mechanisms and even correction and adjustment mechanisms. During the SDRIF debates that took place within the framework of *Workshops*—bringing together councillors, experts and citizens to discuss wide issues or per each main territory in the region—questions were raised that, until then, were relatively unknown in a regional context. These questions were related to the energy crisis, its consequences on the obsolescence of territorial equipment, buildings, services, networks, etc and the increased vulnerability of fragile populations already in very precarious situations. This more particularly provoked their appropriation by political stake-holders, little used to consider ecology as the vector of a new vision of population protection and a guarantor of greater social equality. In this way, a number of innovations were added to the development and monitoring process, even if, from my point of view, they were merely initiated and they merit being amplified and strengthened.

This not the time or place for coming back to the sterile opposition staged over the last few years at top levels of the State for disqualifying planning as a framework tool for the long-term and its integration of scales. I would prefer to come back on a lead initiated by the SDRIF and which could go as far as necessary, as there was no possibility of having it validated by the French Council of State.

To avoid returning to the exercise of the counter-resilient framework paper which claims to immobilize orientations and organizations for the next two decades, complementary acts need to be brought together in a single movement:

---

3. Environmental assessment P. 10.

- Creating a long-term framework for the whole territory.
- Thinking out strategic implementation:
    - Therefore, phrasing and negotiating resources and players who are mobilized in the same sequence.
    - Endowing ourselves with permanent review clauses, places and times for permanently adapting the vision to the reality of progressive contexts.
    - Organizing negotiations with players upstream (from the project's design phase) and implementation so that planning (by a leading player) is linked virtually symbiotically to programming by the partners who are contractually linked by the implementation process.

In fact, this could be one of the solutions for introducing notions of crisis anticipation/ management methods/creation of recovery points and resources into integrated urban construction documents in a coordinated and stabilized way. To enable active resilience to exist on the scale of a regional territory, this domain of work is one of the main domains for planning. It needs to be confirmed.

4 RESILIENCE AS A POLITICAL PROJECT

Resilience, if is a based on a "systemic" mode of urban area planning, most certainly requires for existing administrative and statutory frameworks to be put into question. This leads us to examine questions of governance in two ways:

What governance is needed for a system, which, moreover, is a complete urban system? The question of metropolitan governance is not easy: Are we faced with a new type of situation that is self-regulating and which governs by making all previous forms of greater city governance obsolete? The question is provocative but it does invite political questions, in the wide sense of the term, to be firmly linked with reflections on urban resilience of major metropolis.

4.1 *How do we manage hazards? This requires preparation*

Not only by devices (indicators, instructing parties, leaders, etc … for giving the "right alarm"), but above all by diffusing the new risk culture and educating people on it. The heart of the approach is, in fact, here. This approach is the contrary of a "zero risk" notion. It proposes risk assessment as a foundation linked with democratic debate before the crisis and knowledge of the rules during the crisis. Risk awareness, memories of failures, and risk culture are the foundation stones of resilience: It is not so much the hazard that causes damage, but our incapacity to handle it. As such, "being resilient" would then be the capacity to develop forms of mutualisation and self-organisation. This goes beyond the area of the hazard, the local zone, and also sends us back to the question of global governance of a metropolis. Clearly resilience presupposes a capacity for solidarity between territories.

4.2 *Resilience urges us on to organize geographical and political space and organizations in a suitable way*

To possess the capacity to provide physical assistance to a site that is disorganized by too much or too little water, by a hurricane or by a heat-wave, we will have to think out the ways in which materials will be routed, the flows and networks for going from a preserved area to the point of trauma. This concerns politicians, to delegating public services including adaptations to contractual aspects, to uses … which undoubtedly bring us back to reconsidering the priority geographical areas for town planning. Beside this inventory of what appeared in the SDRIF, and, perhaps, as an introduction to later works, we can make a hypothesis: Is it not possible for new strategic sites of evolution in the Ile-de-France region to be "connectors" for mutualizing means and resources during and after crises caused by hazards?

For this to be the case we will need transparency and knowledge must be circulated. We must learn how to cultivate all the players' and all the citizens' capacity of autonomy and, in the end, develop it. In this way, resilience would become the basis of a specific public policy.

Therefore, from the very beginning, I retain the idea that this notion of resilience—which is so rich—will take on all its meaning if it is extended to new domains: Already very much present in research on risks related to water, already explored in its social meaning and misinterpretation (resilience versus adaptability, a possible social marker by the choice of one term or the other …), it appears to be the bearer of a new "family" of interdisciplinary thought, if only we give ourselves the means for defining it correctly.

Today, organizing moments of interdisciplinary encounters and exchanges of experience on very different territories, leading an increasing number of researchers to concentrate their efforts on resilience, is a necessity for accompanying our change of age and paradigm. Behind the notion of urban resilience, there is a powerful movement transforming our imagination and our moral and political approach to the world.

Consequences of a 100-year flood on the way in which the Paris metropolis and the Ile-de-France region functions:
A few figures on territorial, human and economic issues at stake in flood-prone areas.

Exposure and evolution of territories (46,300 ha potentially exposed –3.85% of the region's total surface)[1]

– The percentage of urbanization on the flood-prone area in Paris and its surrounding "departments" is above 95%.
– In the central city area[2]: 18 200 ha exposed to risks of flooding, including 60% with high to very high risk levels (flooding over 1 metre deep)
– In 2008, areas strictly devoted to individual and collective housing occupy over 6,000 ha, or 13.1% of potentially exposed zones in the Ile-de-France area (46,300 ha). This is the most significant element of urbanization in flood-prone areas, considerably ahead of spaces dedicated to business activities (3,340 ha), facilities (810 ha), transport infrastructures (1,860 ha), or even open urban areas: parks and gardens, sports grounds, golf courses, etc. (5,180 ha).
– Between 1982 and 2008, 1,450 ha of land were urbanized in flood-prone areas, with a significant slow-down over the period 1999–2008, during which urbanization only represents about 200 ha. These areas can be divided up as follows: 47%, in built-up areas, (669 ha), mostly for homes (560 ha), 38% in open urban areas (559 ha) and 15% in transport infrastructures (217 ha).
– Between 1982 and 2008 over 2,600 ha of already built-on land must be added to these 4,500 ha of land urbanized in flood-prone areas. This is land that has evolved (from business activities towards housing, from open areas to facilities, etc.) or which has been renovated, essentially in the heart of the city[3].

Population exposure (830,000 inhabitants directly exposed, or 7.2% of the total regional population) three quarters of which are in the "départements" surrounding Paris)

– Despite the Ile-de-France region's high exposure level, significant numbers of people are continuing to settle in flood-prone areas. If, between 1990 and 1999, exposed populations only progressed by a little under 5 000 inhabitants, between 1999 and 2006, they increased by over 46,000 people, with a growth rhythm (+5.9%) equivalent to that of the region's population (+5.3%) with over half in high risk areas (+15,800 inhabitants.) to very high risk areas (+9,000 inhabitants)[1].
– In the Ile-de-France region, 2/3 of the homes in flood-prone areas are individual detached homes and 1/3 are collective (In the "départements" surrounding Paris,

1/3 of the homes in flood-prone areas are individual detached homes and 2/3 are collective)

- Half of the population in flood-prone areas is exposed to high/very high risk levels (flooding over 1 metre deep).
- In the Ile de France, 100,000 people living in flood-prone areas are also in under-privileged urban neighbourhoods (classified in French regulations as priority urban areas—ZUS (sensitive urban zones), ZRU (urban zones requiring revitalisation), and ZFU (urban free zones).

Economy (56,700 establishments and 630,000 jobs are exposed, or respectively 9.5% and 11.5% of the companies and the numbers of employees listed in the Ile-de-France)[4]

- 85% of the companies in flood-prone areas employ less than 10 people
- Establishments with over 100 employees only represent 1.7% of the number of establishments exposed, but 54% of all employees.
- Between 2000 and 2008, 2.6 million m^2 of the office space made available (2/3 of which were new operations) are located in flood-prone zones (100,000 to 130,000 employees).

---

1. See IAU, Memo N° 557 – July 2011.
2. In the sense of the INSEE definition.
3. See IAU, Memo N° 516 – September 2010.
4. See IAU, Memo N° 534 – February 2011.

Resilience and Urban Risk Management – Serre, Barroca & Laganier (eds)
© 2013 Taylor & Francis Group, London, ISBN 978-0-415-62147-2

Trying to avoid discrepancies between flood victims mobilisations and urban runoffs management policies in order to improve resilience: The case of the Bièvre river catchment

E. Rioust, J.F. Deroubaix & G. Hubert
LEESU, Ecole des Ponts Paris Tech, Université Paris-Est, Marne-la-Vallée, France

Flood risk management can be defined as a complex ensemble of practices, knowledge and regulations tools that aims to improve the protection and/or the functioning of floodable areas, before, during and after the realisation of risk. Consequently, different and numerous types of stakeholders or actors develop particular practices and their own perceptions of how the risk should be managed. In this perspective, numerous works that deals with the resilience insist on the systems' (defined as ensemble of "actors" widely speaking) abilities to anticipate, to live with and to recover from disasters. The resilience concept is more and more often used in risk management (Klein, 2003) because it inherently allows speaking about different types of systems. Some scholars focus on the ecosystems (Holling, 1973), or on socio-ecological systems (Folke, 2006) others on technical systems (De Bruijn, 2005; Lhomme et al, 2010) and others on social and political systems (Handmer et al., 1999; Reghezza-Zitt, 2009; Rioust, 2012). One of the most important aspect of the pretty wide diffusion of the resilience concept is it is used and developed at the same time by different type of scientists and by political institutions (United nations, European Union and also national and local governments).

Within the international documents the notion of resilience is used in the principles of climate change adaptation developed by the International Protocol on Climate Change (McCarthy et al, 2001). The IPCC defines resilience as the ability of organizations to recover a normal condition of functioning after risk realisation. Close to this international community, the International Strategy for Natural Disaster Reduction United Nations (UN ISDR, 2001) proposes to define resilience as follows: *"The capacity of a system, community or society to resist or to change in order that may obtain an acceptable level in functioning and structure. This is determined by the degree to which the social system is capable of organising itself and the ability to increase its capacity for learning and adaptation, including the capacity to recover from a disaster. The motivation to invest in disaster risk reduction is first and foremost a human, people centred concern. It is about improving standards of safety and living conditions with an eye on protection from hazards to increase resilience of communities."* (UN ISDR, 2001).

With these definitions of resilience and adaptive capacities, the individual, communities and their capacity for self-organization are placed at the heart of risk management processes. In this perspective, the international institutions offer a program for risk management, within which the role of communities and individuals is essential to improve safety and "living with risk" standards. Resilience becomes an objective of risk management that has to be added to the former "hazards control" and "vulnerability reduction".

Resilience becomes a political project within which the role of self protection of individuals to natural hazards is emphasized. This political agenda integrates inherently the social construction of risk and thus the capacity of anticipation of human systems in order to suffer damages as less as possible. In this perspective the inhabitants have a role to play in organising crisis management. Resilient behaviours can also be seen as the need to build or to organise waterproof settlements. However, the development of such "resilient" practices depends on the frequency and the gravity of the catastrophe considered. Indeed a high occurrence of risk leads to "accustomed" the inhabitants to risk. Moreover, the inhabitants can organise self-protection and anticipative behaviours only for certain level of risks. In case of very important flooding, the inhabitants can not organise solutions for crisis management and for recovery by themselves. Thus flood management need to be apprehended as an ensemble of practices, or a chain of practices, from which the different actors (inhabitants, professionals of water and sewer services for pluvial floods, elected representatives, and national security services) have all together a key role to play to reduce damages. The resilience literature is not focusing enough on the relations between these different actors and on the perceptions each ones of them develop about the practices of the others.

This proceeding aims to highlight the political dimension of resilience in its implications on the perceptions of the professionals of pluvial flooding risk management. We can noticed that professionals in flooding risk management often assimilate the need to increase resilience to the fact that inhabitants who are living in floodable areas are not aware about the risk. The contents of official documents that deal with resilience, that are insisting on the individuals capacities to react, can also lead to promote this point of view. This perception about the pretended lack of knowledge and lack of awareness about risk is not productive and not founded for the case of Bièvre river catchment.

The Bièvre river catchment is a part of the Paris metropolitan area (figure 1). A wide part of this catchment is highly urbanised and is familiar with pluvial flooding about every two years since 1980's (Bompard, 2010; Rioust, 2012). This type of flood comes from the runoffs of sewerage when the network capacities are exceeded due to heavy rainfall. Fresnes is the city the most exposed to this risk in the catchment because of its situation in the valley and the configuration of sewerage networks (the river has been recovered and separated in different tubes from Fresnes). In Fresnes, certain inhabitants

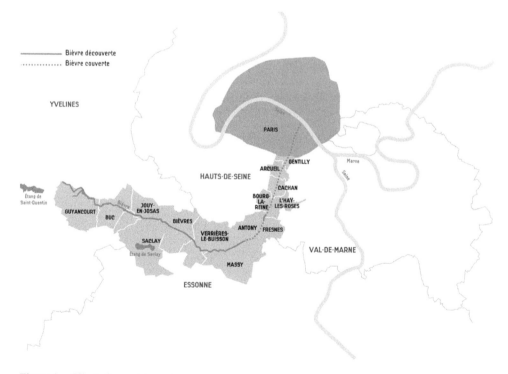

Figure 1.   Bièvre river catchment.

have experienced flooding up to 1, 5 meter inside and outside the houses several times in 1980's, 1990's and most recently in june 2000 and july 2001. Since these events they have known floods in july 2005 and august 2008 (around 20 and 50 cm of sewage inside some houses).

During a 3 years research project about pluvial flooding management we have had the opportunity to conduct interviews with inhabitants living in floodable area in Fresnes. We could have noticed there is discrepancy between the perceptions the professional of urban runoff management developed about the inhabitants "resilience", and the real behaviours of inhabitants. Even if sewer services professionals explain that flooding victims do not know how to react and are definitely not resilient, we could have noticed that victims have knowledge about the risk and develop practices to increase their own protection. As a matter of fact, resilience is a necessity for Fresnes inhabitants and they have not wait for the development of a particulare literature to develop personal measures allowing to increase their ability to react and recover from pluvial flooding.

We are first presenting the different components of Fresnes inhabitants' resilient behaviour before discussing the actual necessity of defining the resilience as a "comprehensive chain of actions" and not only as "vague idea" about how the different actors manage or should manage pluvial flooding risk.

1   FRESNES INHABITANTS' RESILIENCE: KNOWLEDGE, EXPERIENCE AND TECHNIQUES

We have conducted 17 interviews with Fresnes inhabitants to determine how they organise pluvial flooding management. This sample can seemed low but it is important to provide "understanding keys" about real inhabitants' practices. Moreover that kind of data has never been collected for this area. Even if this survey does not concern a large number of floodable inhabitants, it allows catching the reality of "living with risks" inhabitants practices.

Population interviewed counts 6 women and 11 men of different socio economic classes (2 engineers, 1 retired engineer, 1student, 1 lawyer, 1 retired lawyer, 1 retired teacher, 1 mid wife and 1 retired mid wife, 2 social workers, 1 retired social worker, 2 technicians, 1 retired technicians, 2 "others" retired). Only 1 person interviewed is younger than 40 years old, 9 are between 40 and 70 years old, and the others up to 70 years old. The interviews have been gone on different duration between 15 minutes and 2 hours depending on the availability and the will of the inhabitants. These interviews did not aim to produce statistics about inhabitants 'practices and perceptions of flood risk. We wanted to get the reality the inhabitants are living when sewage and pluvial waters are rising up inside the houses and in the streets. For this reason, we had not prepared direct and closed questions. We had let the inhabitants explained their experiences and actions about flood management. This survey shows that interviewed inhabitants are already developing knowledge, experience and techniques to improve·their risk reaction.

The interviews have revelled that inhabitants develop different types of techniques to protect their houses, as well as they have implemented a certain type of alert system. This resilience is competed by the explanations relating to the causes of risk.

Around 80% of interviewed inhabitants have organised individual measures to protect their houses against flooding events. The frequency of this type of flooding led people to organize protective measures, simply because floods are "regulars" and inhabitants are accustomed to it and can build up a certain experience of risk. However it is important to note that interviewed inhabitants were almost the owners of their houses. Thus, the inhabitants have the nessity to organise a way of living with flood because they do not plan to leave this area. We classified individual protection measures into three categories: domestic measures, sealing measures and drainage measures.

Domestic measures correspond to the measures in place inside the house to avoid damage and to facilitate recovery. Mainly, it concerns special layout inside the houses (furniture are elevated, special materials are used to protect the soils and the walls, basements are specially laid out or sealed up) as well as the provision of cleaning materials. More than 30% of interviewed inhabitants have laid out their basements. Domestic measures can also consist in giving the house keys to neighbors in order to let them intervene inside the house in case of floods when owners are absents.

Sealing measures can be presented in two categories. The first refers to techniques designed to prevent sewer overflow inside the houses. It can consist in installing check valve (or non return valve) or a similar "artisanal" technique imagined by inhabitants (positioning a plastic log into the pipe in order to block overflows coming from the sinks or showers pipes) or techniques to block the outlets (sealed manhole covers). The second category refers to measures used to prevent the entrance of the water inside the house through the front door. It consists in building up sheet piling or any kind of "domestic barriers" at the front door. For instance, one can notice that concrete wood or plastic barriers are installed on the doors during the summer. About the choices of techniques implemented, we can notice the importance of the professional culture of local residents. Indeed, people who reported develop techniques that improve the tightness of the house and the drainage of flooded parts have in common to get (or having get) a tehcnical job (engineer, technicians…).

Drainage measures are the installation of pumps in the basement, and the establishment of drains in the basement and in the gardens. These measures are rare in Fresnes certainly because the gravity of floods can not allow developping this measures useful for a low level of water. This idea as been resumed by one inhabitant: *"Anyway, when the water is 50 cm above the sidewalk, you can not do so much. You just have to wait."* (Mrs. L (social worker, aged between 60 and 75), a resident of a street frequently flooded Fresnes "for about 60 years," collected in July 2010).

Almost all the people interviewed in Fresnes have developed techniques that can be qualified as "resilient". These results could conduce to conclude that some discourses about the resilience insist in a too much large extent on techniques and practices already implemented by inhabitants. However inhabitants who develop such measures detain particular social and economic resources. First, these inhabitants own their properties. Secondly, they live in this area for at least 20 years. Thus they have a personal and particular experience of risk. Another aspect of this experience exists through the development of particular alert systems.

2 ALERT SYSTEMS: THANKS TO THE NEIGHBOURS AND THE EXPERIENCE

Pluvial floods are very fast and it is difficult to prevent such events because of the difficulties in monitoring and forecasting storm events. Our survey shows that Fresnes inhabitants alert themselves in case of storm or heavy rainfall events. Their experience of past events and their relations with the neighborhood allows them to detect the "signs" of flood event. The inhabitants have their own "warning indicators," and diffuse the alert through the quarter. These "warning indicators" come from local observations and personal experience. According to inhabitants, they are arlerted by differents elements: local and personal weather forecasting stations, clouds, thunder, alert by the neighbors.

86% of interviewed inhabitants have declared they feel fear when a storm arrives. 25% of interviewed inhabitants say they "go park the car at the top of the street," and 10% "start putting doors barriers at the entrance" when storm arrive. According to the different declarations, only recently settled residents in these areas are not automatically alerted when they know a storm is comming during summer time. But once people experience this type of event, they adopt the same vigilant behavior as their neighbors.

We have not assessed the effectiveness of this local warning systems and we relativize these results because it may be that people "exaggerate" when they deal with their experience of floods. However, these results show that the field experience and some very local observations, make the alert possible and allow inhabitants preparing to face to potential damaging event.

3 NECESSARY RESILIENCE: PROTECTING THE HOUSES, BUT NOT ONLY

Resilience does not consist only in being prepared or to be able to face to risk. Our survey shows that the experience of flood risk create different type of knowledge about the risk. This knowledge encourage the will to act to anticipate the risk as well as the need to explane the causes of risk realisation. According to the interviewed inhabitants, flood events they know are directly related with urban networks runoffs. The people who see overflows from their showers or toilets overflow know that the problem comes from the sewerage networks. It's the same when they see the geysers in the streets from manhole covers.

At Fresnes, 16 of 17 people said that the main cause is the low capacities of sewerage networks. 1/3 of interviewed reported that the only solution capable to protect the quarter is the construction of a water detention tank. However, few people think that the problem is only due to the networ's capacities. For many of them, floods are the result of urbanization and its deconnection with the hydrological characteristics of the area. In Fresnes, some victims also believe that the floods are due to political choices. According to the inhabitants, the elected representatives are not interested in pluvial flooding matters and prefer to deal with, and to engage funds for other type of policies.

We can also notice that only one person Fresnes think floods are caused by exceptional rains. Obviously this question is implied, especially since most of the residents of Fresnes declare they feel fear when a storm comes. But according to inhabitants, it is clear that at Fresnes, the risk is directly related, in their minds, to a network problem or planning decisions. These results lead to define resilience as the abtility for each actor who have responsibilities in risk management to define solutions being aware of what the other actors really do to improve risk management.

4 CONCLUSION

The resilience is closed to the mitigation framework, and consists not so much in protecting from floods but rather in understanding flood risk, preparing for flood risk and living with flood risk. Our investigations in Fresnes have shown that the flood victims have already developed such resilient behaviours because they are "accustomed" to live in a quarters at risk.

People living in this floodable area develop "naturally" protection techniques because it is certainly human to organize self protection when you know your property is at risk. However, the risk culture is not only about houses equipments and techniques. It is the product of an ensemble of knowledge related to personal experience of flood risk and social and economic resources. Individual resilience is not only protecting the houses, it also consist in developing general knowledge about risk and about the real role of each actors who can act to improve risk management.

BIBLIOGRAPHY

Bompard, P. (2010). "Typologie sommaire des inondations dans le Val-de-Marne", Smart Resilience Technology Systems and Tools Conference, Ecole des Ponts Paris Tech.

De Bruijn, K. (2005). "Resilience and flood risk management: A systems approach applied to lowland rivers".

Folke, C. (2006). "Resilience: The emergence of a perspective for social-ecological systems analyses". Global environmental change, 16(3), pp. 253–267.

Handmer, J.; Dovers, S.; and Downing, T. (1999). "Societal vulnerability to climate change and variability". Mitigation and adaptation strategies for global change, 4(3), pp. 267–281.

Holling, C. (1973). "Resilience and stability of ecological systems". Annual review of ecology and systematics, 4, pp. 1–23.

Klein, R.; Nicholls, R.; and Thomalla, F. (2003). "Resilience to natural hazards: How useful is this concept?". Global Environmental Change Part B: Environmental Hazards, 5(1–2), pp. 35–45.

Lhomme, S.; Serre, D.; Diab, Y.; and Laganier, R. (2010). "Les réseaux techniques face aux inondations ou comment définir des indicateurs de performance de ces réseaux pour évaluer la résilience urbaine". Bulletin de l'Association de géographes français. Geographies, 487, pp. 502.

McCarthy, J. (2001). Climate change 2001: impacts, adaptation, and vulnerability: Contribution of Working Group II to the third assessment report of the Intergovernmental Panel on Climate Change. Cambridge University Press.

ONERC (2011). Plan National d'Adaptation au Changement Climatique. Ministère de l'Ecologie, de l'Energie, du Développement Durable et de la Mer.

Reghezza-Zitt, M. (2009), "Introduction au séminaire résilience urbaine". In Conférence à l'Ecole Nationale Supérieure de Géographie, Paris.

Rioust E. (2012), "Gouverner l'incertain: adaptation, résilience et évolutions dans la gestion du risque d'inondation urbaine. Les services d'assainissement de la Seine-Saint-Denis et du Val-de-Marne face au changement climatique." PhD Ecole des Ponts Paris Tech, Université Paris-Est.

Timmerman, P. (1981). "Vulnerability, resilience, and the collapse of society". Environmental Monograph, 1.

UN/ISDR (2001), Report of Working Group 3 to the ISDR Inter Agency Task Force for Disaster Reduction. UN/ISDR, Geneva.

Author index